# Electrocatalysis and Electrocatalysts for a Cleaner Environment - Fundamentals and Applications

*Edited by Lindiwe Eudora Khotseng*

Published in London, United Kingdom

IntechOpen

*Supporting open minds since 2005*

Electrocatalysis and Electrocatalysts for a Cleaner Environment - Fundamentals and Applications
http://dx.doi.org/10.5772/intechopen.91567
Edited by Lindiwe Eudora Khotseng

Contributors
M. Amin Farkhondehfal, Juqin Zeng, Mohammadali Beheshti, Saeid Kakooei, Mokhtar Che Ismail, Shohreh Shahrestani, Nisar Ahmed, Guilherme M. Martins, Najoua Sbei, Geórgia C. Zimmer, Lindiwe Eudora Khotseng, Adebare Nurudeen Adewunmi, Sabeje Akindeji Jerome, Su Huaneng, Akhoury Kumar Sudhir Kumar Sinha, Umaprasana Ojha, Salah Kouass, Hassouna Dhaouadi, Abdelhak Othmani, Fathi Touati, Phuoc Huu Le, Dang Long Quan, Qian Xu, Jiajia Zhang, Chunzhen Yang



First published in London, United Kingdom, 2022 by IntechOpen
IntechOpen is the global imprint of INTECHOPEN LIMITED, registered in England and Wales, registration number: 11086078, 5 Princes Gate Court, London, SW7 2QJ, United Kingdom
Printed in Croatia

British Library Cataloguing-in-Publication Data
A catalogue record for this book is available from the British Library

Additional hard and PDF copies can be obtained from orders@intechopen.com

Electrocatalysis and Electrocatalysts for a Cleaner Environment - Fundamentals and Applications
Edited by Lindiwe Eudora Khotseng
p. cm.
Print ISBN 978-1-83968-127-1
Online ISBN 978-1-83968-128-8
eBook (PDF) ISBN 978-1-83968-129-5

For EU product safety concerns:
IN TECH d.o.o., Prolaz Marije Krucifikse Kozulić 3, 51000 Rijeka, Croatia, info@intechopen.com or visit our website at intechopen.com.

# Meet the editor

Lindiwe Khotseng has been an associate professor in the Chemistry Department, University of the Western Cape, South Africa since 2015. Prior to this, she worked as the managing director of ESKOM Centre for Electrocatalytic Research, South African Institute for Materials Research, University of the Western Cape. Dr. Khotseng obtained her Ph.D. in Chemistry from the University of Stellenbosch, South Africa, in 2004 and PG Diploma in T&L (HE) at the University of the Western Cape in 2022. Her research focus is on energy and nanomaterials. She has published in qualified international journals or specialized books and presented at international chemical meetings. She has graduated sixteen MSc and five Ph.D. students. She has served on various leadership committees at her faculty and NRF panels. She received the DTI Technology Award as Outstanding Black Researcher in her research field in 2013. Dr. Khotseng is a professional member of the South African Chemical Institute (SACI), the Organization for Women in Science for the Developing World (OWSD), and the International Society of Electrochemistry (ISE).

# Contents

# Preface

Over the past decade, electrocatalysis scientists have been at the forefront in addressing various challenges in climate change, pollution, and sustainable energy. Electrocatalysis is instrumental for addressing global challenges in energy and the environment at a high level. Electrocatalysts are the type of catalysts that are used in electrochemical processes/electrocatalysis.

This book discusses electrocatalysis and electrocatalysts for energy, water electrolysis, water treatment, $CO_2$ conversion, and green chemistry. It reviews various electrocatalysts and their characterization, properties, and electrochemical performances.

The first section of the book covers topics in direct alcohol fuel cells including Pt-based electrocatalysts as non-carbon electrode support materials and the development of electrocatalysts for direct methanol fuel cells.

The second section of the book covers various topics in electrocatalysis and electrocatalysts for a cleaner environment. Chapters address such topics as electrocatalysts for the conversion of $CO_2$ to valuable products and SYNGAS, electrocatalysts for water electrolysis, functionalization of C-H groups through electrocatalysis processes that offer green chemistry and possibility of scale-up with improved reaction times, kinetics and yields, and the characterization, properties and electrochemical performances of photocatalysts in electrochemical processes.

**Lindiwe Eudora Khotseng**
Department of Chemistry,
University of the Western Cape,
Cape Town, South Africa

Section 1

# Electrocatalysts for Energy/ Fuel Cell Applications

Chapter 1

# Recent Advances in Pt-Based Binary and Ternary Alloy Electrocatalysts for Direct Methanol Fuel Cells

*Dang Long Quan and Phuoc Huu Le*

## Abstract

The direct methanol fuel cell (DMFC) is among the most promising alternative energy sources for the near future owing to its advantages of simple construction, compact design, high energy density, and relatively high energy-conversion efficiency. Typically, the electrodes in DMFC is comprised of a Pt-based catalysts supported on great potential of carbon materials such as multi-walled carbon nanotubes (MWCNTs), carbon black (CB), graphene, etc. It is desired to develop an electrode with high surface area, good electrical conductivity and suitable porosity to allow good reactant flux and high stability in the fuel cell environment. This chapter will provide recent advances in Pt-based binary and ternary electrocatalysts on carbon supports for high-performance anodes in DMFC. Through studying the effects of composition-, support-, and shape dependent electrocatalysts, further fundamental understanding and mechanism in the development of anode catalysts for DMFC will be provided in details.

**Keywords:** direct methanol fuel cell (DMFC), electrocatalysts, Pt-based binary and ternary alloys, carbon based supports

## 1. Introduction

### 1.1 Structure and operating principles of direct methanol fuel cell

Today, population growth and economic growth lead to an increased energy demand. Current energy sources are mainly from fossil fuels (coal, oil and gas), which produce carbon dioxide ($CO_2$) and other greenhouse gases that are responsible for global climate change. In order to minimize the bad impact of greenhouse effects (e.g. acid rain, ozone damage), the world needs an appropriate transition of the energy sources being used. Therefore, the development of clean energy is a common concern for balancing economic, social development, and environmental protection. Fuel cells are one of the most promising energy sources for use in transportation and communication applications. Compared with internal combustion engine, fuel cells are environmentally friendly, durable, reducing noise, and so on [1]. Currently, the main fuel cells include alkaline fuel cell (AFC), polymer electrolyte membrane fuel cell (PEMFC), direct methanol fuel cell (DMFC), phosphoric acid fuel cell (PAFC) and solid oxide fuel cell (SOFC).

Among the fuel cells, DMFC, apart from the advantage of being environmentally friendly, also has a high energy density [2, 3]. DMFC is one of the popular types of fuel cells using methanol directly as fuel. DMFC with liquid fuel can operate at ambient air temperature, has good energy density and is easy to store and transport. The membrane electrode assembly with acid or alkaline membranes is a main component of DMFC, in which both side of the polymer electrolyte membrane contact to anode and cathode catalyst layers. Conventionally, PtRu/C or PtRu catalyst is used in the anode, while Pt/C or Pt in the cathode [4]. The gas diffusion layers are closely aligned with the catalyst layers to aid reactant distribution, current collection and catalytic protection.

In DMFC, an electrochemical reaction will occur at the anode due to the interaction between methanol and water to produce protons and electrons. Specifically, at the anode, six protons and six electrons are formed by a methanol molecule reacting to a water molecule. These protons can move freely through the electrolyte toward the cathode, while electrons can travel through the external load (**Figure 1**). In addition, carbon dioxide will be also formed by oxidization of methanol. Meanwhile, at the cathode, water is formed by oxygen electrocatalytic reduction reaction. Therefore, the number of electrons at the anode is larger than the number of electrons at the cathode, resulting in a potential established between the two electrodes. Reaction formulas in DMFC are shown as follows [5]:

$$\text{Anode reaction}: CH_3OH + H_2O \rightarrow CO_2 + 6H^+ + 6e^- \quad (1)$$

$$\text{Cathode reaction}: 3/2\,O_2 + 6H^+ + 6e^- \rightarrow 3H_2O \quad (2)$$

$$\text{Overall reaction}: CH_3OH + 3/2\,O_2 \rightarrow CO_2 + 2H_2O \quad (3)$$

### 1.2 Electrocatalytic materials in DMFC

In the early 1950s, the anode and cathode electrocatalysts used for methanol fuel cell began to be investigated. Initially, methanol fuel cell used an alkaline electrolyte with an anode catalyst of nickel or platinum for methanol electro-oxidation reaction, and silver for the oxygen reduction process. At the same time, studies of acidic electrolyte replacement have shown that the kinetics of methanol electro-oxidation are slower in this environment than in alkaline [6]. However, DMFC using liquid alkaline electrolyte has a main drawback of carbonate formation, meanwhile, DMFC using an acid electrolyte presents better perspectives. The Pt-Sn bimetallic catalyst has been systematically studied by Jansen and Molhuysen [7], which promoted the use of bimetallic catalysts for DMFC. Along with Pt-Sn, Pt-Ru was the most potential bimetallic catalyst for anode formulations, but it was still underestimated compared with Pt-Sn bimetallic catalyst.

During the 1960s, Pt-Ru system, particularly Pt combining Ru in solid solution, revealed great potential applications supported by the studies of Watanabe and Motoo [8]. In the 1960s and 1970s, the study of anode's processes was carried out by many different groups through the search or improvement of a suitable catalyst as a premise for the construction of the bifunction theory for bimetallic catalysts based on methanol oxidation. In twenty years later, the structural, surface and electronic properties of the most promising systems for DMFC, essentially Pt-Ru were investigated. Besides, the studies of electrode structure including diffusion and backing layers also attracted a lot of attention. Most of these studies aimed to enhance the catalytic activity, improve reaction rate, and minimize poisoning due to methanol

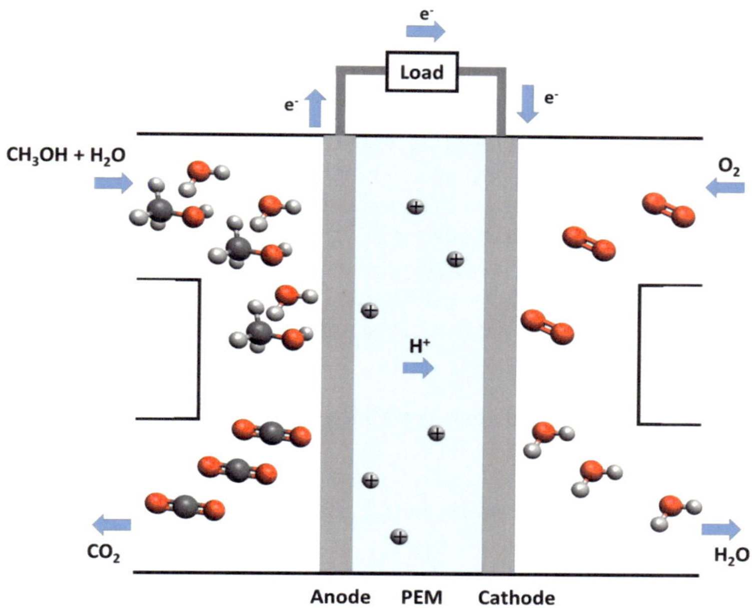


**Figure 1.**
*Structure and operating principle of DMFC.*

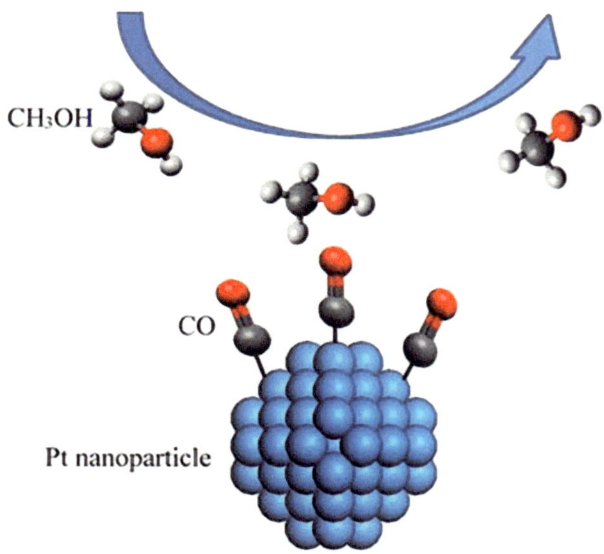


**Figure 2.**
*Poisoning CO on Pt surface to prevent methanol oxidation catalysis.*

residues by combining different metals with platinum. It has been found that the use of metal alloys can modify electronic surface structure, physical structure to prevent CO poisoning and absorb oxygen/hydroxyl species. The 1990s marked significant advances in DMFC technology with early applications for portable electronic devices. Briefly, fuel cells (including DMFC) were widely studied in the early 20th century.

DMFC is one of the most potential candidate of fuel cells, however, slow electro-oxidation kinetics, methanol crossover, and gas management on the anode side in DMFC need to be improved. In methanol electro-oxidation, various surface intermediates as CO, $COH_{ads}$, $HCO_{ads}$, $HCOO_{ads}$ are formed and strongly adsorbed to the surfaces of catalysts. As a result, methanol molecules are prevented from the next actions, leading to slow down the oxidation reaction [5]. In addition, fuel efficiency is decreased because of small percentage of the intermediates desorbing before being oxidized to $CO_2$. Therefore, research on developing suitable catalysts to prevent CO poisoning and improve efficiency for DMFC is one of the critical issues. One of the most popular intermediates, carbon monoxide was produced by adsorption and de-protonation on the anode catalyst, which limits the rate of methanol oxidation. Specifically, if CHO or COH are directly dehydrogenated, carbon monoxide will be formed, as shown in **Figure 2**. Consequently, the active sites of the catalyst will be decreased that limits the next reactions to be occurred.

## 2. The binary Pt-based and ternary PtRuM alloy electrocatalysts for DMFC

### 2.1 The binary Pt-based alloy electrocatalysts for DMFC

#### *2.1.1 Advantages of binary Pt-based alloy catalysts compare to platinum catalyst*

Around the middle of the nineteenth century, the first model of a fuel cell using a platinum wire as an electrode was investigated [4]. Since then, the application of platinum in fuel cells is great of interest for scientists. In strongly acidic electrolytes, platinum nanoparticles are higher catalytic activity and better stability than other noble metals. Therefore, it was used as an optimal catalyst in the first most studies of DMFC. However, platinum's cost is high and easy to dissolve and agglomerate under poor operating conditions, resulting in the commercialization of DMFC based on platinum catalyst is still limited. Especially, Pt-based anode can be poisoned by a carbonaceous intermediate as CO, $COH_{ads}$, $HCO_{ads}$, $HCOO_{ads}$, leading to decreasing DMFC efficiency. To overcome the above obstacles, many studies have been conducted to improve catalysts in DMFC. As a result, binary catalysts are one of the most effective catalysts, in which Pt-Ru system is the most common.

According to the bifunctional mechanism, each metal in surface of Pt-M will promote various steps in the overall reaction. The Pt – CO strength bond and the reaction activation energy can be modified by energy perturbations of the surface d-band of Pt due to ligand effect from M metal. As a result, CO on the platinum surface requires less energy to oxidize to $CO_2$ compared with pure platinum. In addition, the attachment of metal M on the substrate can induce changes in the catalytic properties of the substrate, which is attributed to establishing a new equilibrium state for occurring the electronic and strain effects simultaneously. The presence of metal M significantly reduces the onset potential for the reaction because the oxidation of CO on the surface with more metal M is greatly affected compared to pure Pt.

Therefore, bimetallic catalysts have been developed over many decades. Specifically, the combination of platinum with transition metals (Fe, Co, Ni) not only improves electrochemical performance but also decreases the Pt mass loading and enhances methanol tolerance in the oxygen reduction reaction (ORR) [9–11]. ORR activity and stability of the Pt catalyst can be increased in the presence of Cr and Pd. In addition, the bimetallic catalyst offers advantages to the reactions occurring at the DMFC electrode with outstanding features such as improving methanol dissociative chemisorption, reducing CO poison adsorption and CO removal via its oxidation by adsorbed OH [12–14].

Survey results of methanol oxidation ability by cyclic voltammetry (CV) measurement showed that the methanol oxidation capacity of all Pt-M alloys was superior to that of pure Pt. Many research groups have in turn fabricated the Pt-M alloys such as PtFe, PtCo, PtNi, PtCu, PtMo, PtRu, PtRh, PtPd, etc., and investigated their activity to oxidize methanol comparing to pure Pt catalyst [15–22]. The obtained CV spectra results are shown in **Table 1** and **Figure 3**. Clearly, although there were differences in the selection of substrate, the composition percentage in the alloy, or the concentration of the investigated solution, the ability to oxidize methanol of pure Pt is always lower than that of Pt-M alloys. Here, in the CV results, methanol oxidation capacity is realized by the spectral peak position during the forward sweeping/positive sweeping (maximum current density - $j_f$). The value of $j_f$ in CV spectrum of PtFe, PtCo, PtNi, PtCu, PtMo, PtRu, PtRh, PtPd alloys and pure Pt are listed in **Table 1**. It can be clearly seen that the $j_f$ values of PtFe, PtNi, PtCu, and PtRu alloys were about twice that of pure Pt. The difference is about 3–5 times for Pt67% Mo33% and Pt54% Rh46%. Especially, PtCo (1: 9) and Pt3Pd1 offered the enhancement of ~10 times. In adition, the CV spectra of PtRu alloys almost had no current density peak for reverse sweeping. This indicates that the high tolerance to CO poisoning led to the increased efficiency of methanol oxidation and the enhanced DMFC performance. Furthermore, the CV results of the PtCo, PtMo, PtRh and PtPd alloys show that the ability to oxidize methanol was also strongly influenced by the composition ratios in the alloy. It is clear to conclude that

| **Pt alloy** | **Support** | **Size (nm)** | **Electrooxidation** | | | **Reference** |
|---|---|---|---|---|---|---|
| | | | **Condition** | **Current density of Pt (mA cm$^{-2}$)** | **Current density of PtM (mA cm$^{-2}$)** | |
| PtFe | Multi-walled carbon nanotubes (MWCNT) | 2 | 0.5 M $H_2SO_4$ + 1 M $CH_3OH$ | 0.12 | 0.24 | [15] |
| PtCo | Graphene oxide (GO) | 2.1–3.4 | 1 M $H_2SO_4$ + 2 M $CH_3OH$ | 3.85 | 38 PtCo (1:9) | [16] |
| PtNi | Functionalized carbon nanotubes (FCNTs) | 2.7–3.9 | 0.5 M $H_2SO_4$ + 1 M $CH_3OH$ | 300 | 500 | [17] |
| PtCu | Graphene oxide (GO) | 3 | 0.5 M $H_2SO_4$ + 0.5 M $CH_3OH$ | 0.83 | 1.6 | [18] |
| PtMo | Single-wall carbon nanotubes (SWCNT) | — | 0.5 M $H_2SO_4$ + 0.5 M $CH_3OH$ | 0.1 | 0.55 Pt67%Mo33% | [19] |
| PtRu | Carbon (C) | 4.5 | 1 M $H_2SO_4$ + 2 M $CH_3OH$ | 7.5 | 17.5 | [20] |
| PtRh | — | 4.3–6 | 0.5 M $H_2SO_4$ + 1 M $CH_3OH$ | 140.7 | 404.9 Pt54%Rh46% | [21] |
| PtPd | Vulcan XC-72R | 6.5 | 0.5 M $H_2SO_4$ + 0.5 M $CH_3OH$ $N_2$-saturated | 25 | 200 Pt3Pd1 | [22] |

**Table 1.**
*Various studies of advantages of binary Pt-based alloy compare to platinum catalyst.*

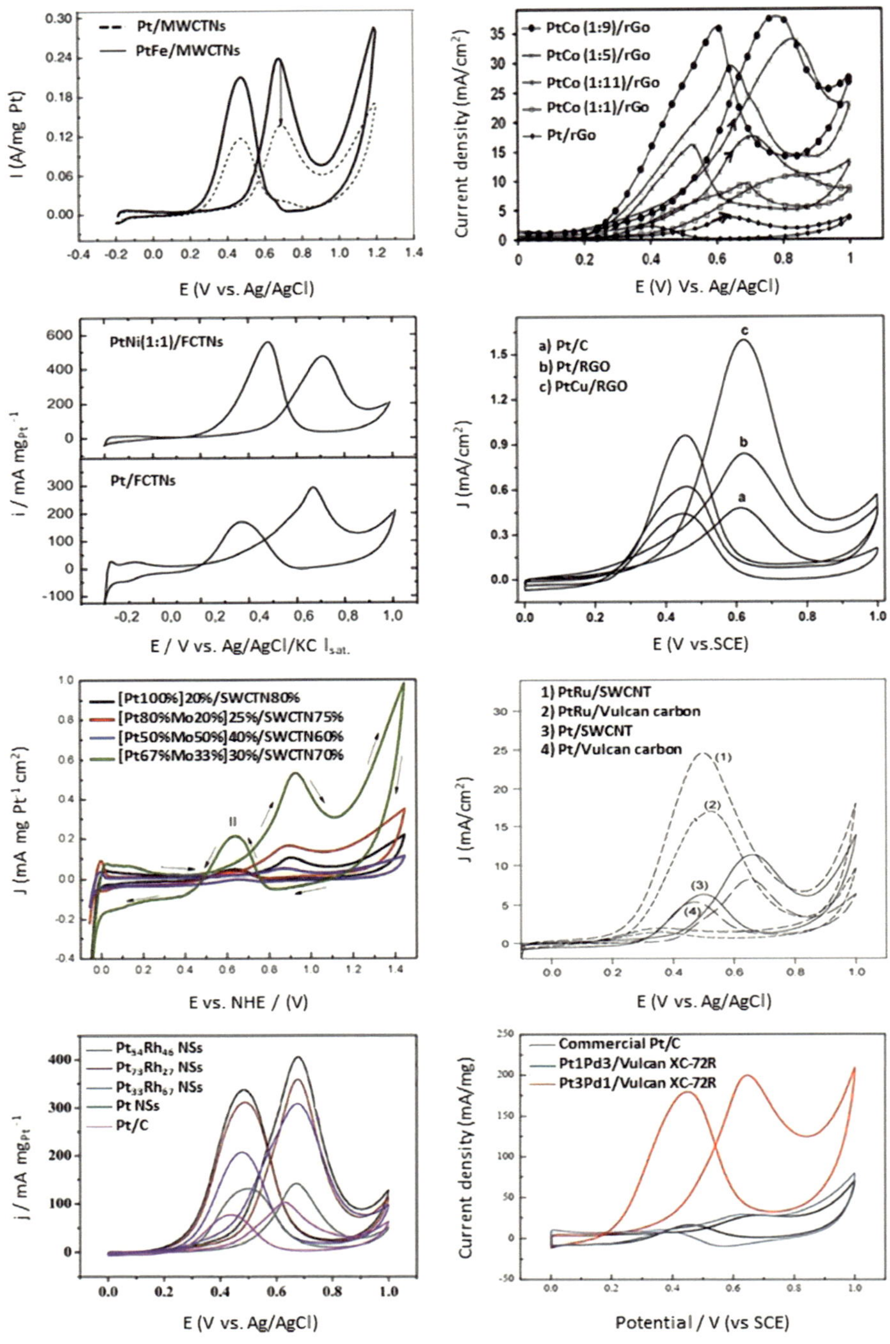

**Figure 3.**
*Results of methanol oxidation ability by CV spectra of PtFe/MWCNTs, PtCo/rGO, PtNi/FCNTs, PtCu/RGO, PtMo/SWCNT, PtRu/Vulcan carbon, PtRh, and PtPd/Vulcan XC-72R in comparison with pure Pt [15–22].*

Pt-M alloys exhibited higher methanol oxidation performance than that of pure Pt. However, it is not suitable to make a comparison among the different alloys because some alloys require careful examination of the composition ratios. Also, it is worthy of note that the differences in the onset potential position of $j_f$, the $j_f/j_r$ ratio, and the substrate can influence the evaluation of methanol oxidation performance.

#### 2.1.2 Dominances of PtRu catalyst compare to other binary Pt-based alloy catalysts

Pt-Ru has been interested in research since the late 19th century with the first mention in thermoelectric materials. In 1930s, the physical and chemical properties of this material system were explored, setting the stage for more detailed studies in the following decades.

First, the Pt-Ru system is one of the most common alloys used for fuel cells owing to its lower potential onset than Pt or Pt alloys. Indeed, **Table 2** shows that the strongest synergetic effects belonged to Pt-Ru. Specifically, their onset potentials in 1 M $H_2SO_4$ at 100°C and current density 20 mA/cm$^2$ about 0.24–0.32 V, while this value of Pt is 0.44 V, and Pt-M (M belongs to group IVB, VB, VIB, VIIB and some other metals) have the onset potentials in range of 0.3–0.44 V. The electrolytic activity of Pt and its alloy related to the oxidation of methanol has been established the foundation for further studies. The extensive studies of combining Pt with multiple elements in the periodic table to find the most suitable catalysts (including Pt-Ru) for organic fuel oxidation were also carried out in the 1960s by Batelle group [24].

Second, the Pt-Ru system is often used for fuel cell catalysis because Ru and its alloy can be used as an effective catalyst. In parallel with Pt-Ru studies, the other Ru systems (i.e. Ru-Ta, Ru-Ni, Ru-Rh and Ru-Ir) were also investigated. Corrosion and electrochemical properties were strongly influenced in Ru systems doped with Pb, Tl, Ag [24]. A ruthenium electrode with a high surface area was successfully synthesized by a research team from Moscow [24], and the effect of heat treatment on ruthenium dispersion was also studied. In addition, the effect of temperature on ruthenium properties was also studied [24]. The calculation of electronic structures combined with density function theory was used to investigate the free-energy landscape of the electrochemical oxygen reduction reaction over Pt (111), in which the changes of applied bias confirmed that noble metals were suitable for the peroxide mechanism [25]. In this mechanism, the oxygen reduction on the surface of Pt or noble metals is performed through peroxy intermediates according to the following scheme:

$$O_2 + {}^* \rightarrow O_2{}^* \tag{4}$$

$$O_2{}^* + \left(H^+ + e^-\right) \rightarrow HO_2{}^* \tag{5}$$

$$HO_2{}^* + \left(H^+ + e^-\right) \rightarrow H_2O + O^* \tag{6}$$

$$O^* + \left(H^+ + e^-\right) \rightarrow HO^* \tag{7}$$

$$HO^* + \left(H^+ + e^-\right) \rightarrow H_2O + {}^* \tag{8}$$

Here, the "*" symbol indicates a site on the surface. This is an important basis for the application of Pt and noble metals for fuel cell catalysis.

Finally, the strain effect caused by the lattice mismatch between platinum and ruthenium reduced the binding energy of intermediates such as carbon monoxide, resulting in the increased tolerance to CO poisoning for PtRu bimetallic [26, 27]. Bifunctional theory of the electrocatalysis was formulated based on an investigation of Pt-Ru activity by Watanabe and Motoo [26, 27]. According to this mechanism,

| Catalyst | The onset potential at 100°C at 20 $mA.cm^{-2}$ (V) |
|---|---|
| Pt | 0.44 |
| PtM<br>M in group IVB and VB | 0.37–0.39 |
| PtM<br>M in group VIB and VIIB | 0.20–0.38 |
| PtM<br>M: Fe, Co, Ni, Be, Pb, Cu, Bi, Sb | ≥ 0.44 |
| **PtRu** | **0.24–0.32** |
| PtOs | 0.30–0.35 |
| PtRh | 0.44 |
| PtIr | 0.33 |

**Table 2.**
*The catalytic activity of Pt and Pt alloys in methanol oxidation [23].*

generation of active oxygen species belongs to ruthenium, while the adsorbed methanol species are kept by platinum. Hence, methanol oxidation takes place by the following reactions [28]:

$$Pt + CH_3OH \rightarrow Pt - CH_3OH_{ads} \rightarrow Pt - COH_{ads} + 3H^+ + 3e^- \quad (9)$$

$$Pt - COH_{ads} \rightarrow Pt - CO_{ads} + H^+ + e^- \quad (10)$$

$$Ru + H_2O \rightarrow Ru - OH_{ads} + H^+ + e^- \quad (11)$$

$$Pt - COH_{ads} + Ru - OH_{ads} \rightarrow Pt + Ru + CO_2 + 2H^+ + 2e^- \quad (12)$$

$$Pt - CO_{ads} + Ru - OH_{ads} \rightarrow Pt + Ru + CO_2 + H^+ + e^- \quad (13)$$

The invent of the bimechanism has underpinned the profound and extensive studies of the Pt-Ru catalyst for DMFC.

## 2.2 The ternary PtRuM electrocatalysts for DMFC

From the above analysis, PtRu is the most commonly alloy for the anode methanol oxidation electrocatalyst in DMFC. However, both platinum and ruthenium are precious metals, so in order to increase performance and reduce costs, the studies of searching effective ternary PtRuM electrocatalysts have been attracting a lot of attention. It has been found that ternary Pt alloys can improve catalytic performance as compared to Pt or PtRu because the additional metals can change the electronic properties or the surface structure of the Pt or PtRu [29]. Indeed, from theoretical screening results, Strasser et al. showed that ternary PtRuM (M = Fe, Co, Ni, Rh, Ir) which had better CO-tolerance than pure Pt or PtRu to result in a superior catalytic activity [30]. This is the key platform for driving the fabrication of ternary catalysts toward improving the efficiency and reducing the cost of DMFC.

Due to some outstanding properties such as low potential onset, CO-tolerance, and ability to participate in oxygen reduction reaction at a fuel cell cathode, molybdenum (Mo) is also one of popular elements of interest for the development of DMFC. Binary PtRu has been successfully formulated and exhibited the better tolerate CO poisoning than pure Pt [31–34]. Differ from PtRu, the main oxidation peak of PtMo shifts toward a lower potential, while the oxidation of CO in PtMo occurs in two steps to form two peaks in the voltamperometric sample [31]. It was found that Mo species also reduced easily due to the undergo oxygen transfer. In addition, CO oxidized by Mo at a low voltage had weakly bonding, and thus the CO occupied a very low proportion of the total CO absorbed to the catalyst [31–35]. Therefore, PtMo has a lower methanol oxidation activity than PtRu. However, the two different CO oxidation mechanisms in PtRu and PtMo at different potentials can induce a co-catalytic effect by combining Ru and Mo to form a ternary PtRuMo alloy [36]. This was the basis for a series of studies on ternary PtRuMo catalysts. Specifically, PtRuMo/C was prepared by a two-steps reduction method, and PtRuMo/C possessed a decrease of CO poisoning as compared to PtRu and Pt [37]. Moreover, DMFC with PtRuMo/CNTs anode had higher performance than DMFC with PtRu/CNTs or PtRu/C anodes [38]. The successful synthesis of PtRuMo nanoparticles on MWCNTs by chemical reduction under hydrothermal synthesis method, and the PtRuMo/MWCNTs exhibited superior catalytic activity and durability for methanol oxidation in $H_2SO_4$ solution over PtRu/MWCNTs [39]. The CV and electrochemical impedance spectroscopy (EIS) results for $Pt_{43}Ru_{43}Mo_{14}$/MWCNTs, $Pt_{50}Ru_{50}$/MWCNTs, and Pt/MWCNTs presented that $Pt_{43}Ru_{43}Mo_{14}$/MWCNTs obtained the highest activity and stability [40].

Iron (Fe) is also one of the good candidates for electrode catalysis in DMFC anode. The presence of iron as the third element in the platinum alloy can weaken the Pt-CO bonding [41, 42]. It is known that, if platinum and iron combine, the orbital mixing will occur. Due to electron-rich platinum and electron-poor iron, Pt electron density and electron density hinders the electron back-donation from Pt to CO to result in weakening Pt-CO bonding. In addition, the cost of iron-containing precursors is much lower than that of Pt and Ru. Therefore, the cost of DMFC using PtRuFe catalyst will significantly reduce, and thus expand the commercialization capability of the DMFC. For these reasons, the study of PtRuFe alloys was promoted. In particular, PtRuFe/C (2:1:1 atomic ratio, 60 wt% metal) was synthesized by impregnation method [43]. **Figure 4** shows that the PtRuFe nanoparticles were uniformly formed with an average size of approximately 2.5 nm. For CO-stripping results, Jeon et al. [43] obtained the onset potentials of PtRuFe/C and PtRu/C were 0.44 V and 0.49 V, respectively. In addition, current intensity of PtRuFe/C was also higher than that of PtRu/C under the similar experimental conditions. The obtained results confirmed that PtRuFe/C exhibited the better methanol electro-oxidation activity [43]. In another study, PtRuFe nanodendrites which was prepared via a one-pot solvothermal method also presented the higher specific and mass activities than those of PtFe and Pt [44].

In order to minimize CO poisoning and increase the performance of the DMFC, ternary catalyst including platinum, ruthenium and cobalt is also considered. The PtRuCo/GC catalysts was successfully prepared by the one-step electrochemical CV co-deposition method, and it exhibited higher electrochemical activity and stability for methanol oxidation than PtRu/GC [45]. PtRuCo/C was successfully synthesized by the electro-deposition process with controlling the deposition potential and deposition time to control Pt and Ru galvanic displacements [46]. It was found that PtRuCo/C presented superior catalytic activity and tolerance to CO poisoning as compared to those of the commercial PtRu/C [46].

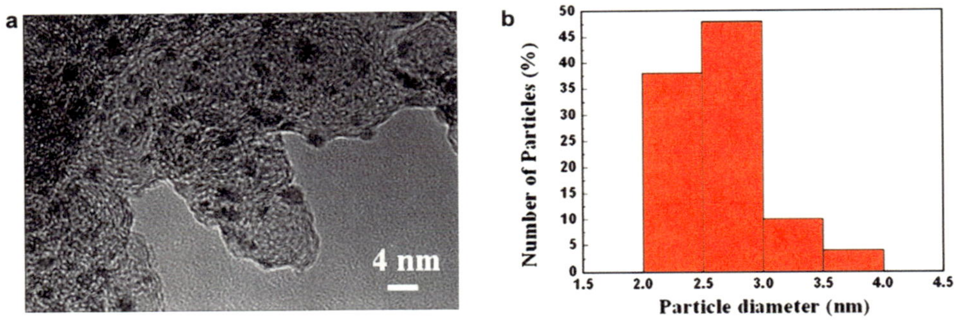


**Figure 4.**
*A TEM image (a) and particle size distribution (b) of PtRuFe/C (2:1:1) catalyst [43].*

The theoretical investigation found that the segregation processes for creating Pt surface enrichment was unlikely to take place in the PtNi system [47]. Additionally, unlike Ru in PtRu, Ni in PtNi was not dissolved in the electrolyte due to the nickel-hydroxide-passivated surface and the increased stability of Ni in the Pt network [48]. Furthermore, the use of Ni as the third metal in a ternary alloy will reduce the amount of noble metals of Pt and Ru that allows reducing the DMFC cost and supports for its commercialization. Therefore, ternary PtRuNi alloys have been developed and characterized [49–51]. Specifically, PtRuNi/C catalyst can significantly improve the methanol oxidation performance compared with PtRu/C [49–51]. Liang et al. were successfully synthesized a carbon-supported PtRuNi nanocomposite via a microwave-irradiated polyol plus annealing synthesis strategy, and the nanocomposite exhibited an enhanced tolerance to CO [52]. The impedance patterns of methanol electrooxidation for PtRuNi/C proved its superior performance to that of PtRu/C [53]. Similar results were also obtained for PtRuNi and PtRu on CNTs [54]. Adding a third metal to PtRu is one of the effective approaches to improve DMFC performance. Therefore, in addition to the aforementioned three-component alloys, the other three-component alloys such as PtRuCu, PtRuW, PtRuMn, etc. were also investigated in the literature [55–57].

Briefly, the ternary platinum alloy exhibited superiority in catalytic activity and resistance to CO compared with bimetallic or pure metals. It is worthy to note that each type of catalytic substrate also provides certain advantages for the DMFC application. Therefore, it is essential to study ternary platinum alloy on different catalytic supports to look for the optimal material system with excellent DMFC performance and cost effectiveness.

## 3. Effects of supports, shape, and structure on methanol oxidation activity of catalysts

### 3.1 Effects of supports on methanol oxidation activity of catalysts

In the early stages of fuel cell development, only Pt nanoparticles were used as catalysts for the anodes and cathodes. The Pt catalyst showed higher activity and durability than any other metals [58]. However, without a substrate, the increased Pt concentration leads to the agglutination of the Pt nanoparticles, and consequently reduces the surface area, which in turn affects the catalytic performance [59]. Therefore, the use of carbon-based materials (i.e. carbon black (CB), carbon nanotubes (CNTs), carbon nanofibers (CNFs), and graphene, etc.) allows reducing the amount of Pt used, while increasing the activity of the electrolytes.

Carbon materials with high surface area, suitable porosity, high corrosion resistance, and excellent electrical conductivity, allow for multiple dispersing of Pt nanoparticles and accelerate electronically charge transfer from electrode to membrane electrolysis [60]. Oxide materials are also good candidates for serving as the substrates of Pt catalyts because they can improve corrosion resistance to carbon materials [61, 62]. Since the 1990s, CB materials have been widely used as catalysts for Pt and its alloys in PEMFC because of the large area surface area, good electrical conductivity, porous structure, and low cost. Therefore, CB has been widely used for many decades as a base material for fuel cell catalysis [63]. Under the normal conditions, CNTs has the same advantages as CB, but it supports for minimizing the agglomeration of Pt nanoparticles. Specifically, Shao et al. found that Pt/CNTs were more stable than Pt/C for PEMFC primarily because the impedance of CNTs was higher than CB [64]. Furthermore, CNTs appears to be a good support material for fuel cells because of their high electrical conductivity, purity, and durability compared with the conventional substrates [65–72]. It is well-known that the main component of DMFC is the membrane electrode assembly (MEA), where the anode and cathode reactions occur to generate electrical energy. The ideal MEA material has the following prerequisites: (1) An efficient anode catalyst is for oxidation of methanol completely, (2) solid polymer electrolyte has a high proton conductivity and low methanol permeability, (3) the cathode catalyst is methanol-tolerant with a high deoxygenation activity [73]. Therefore, it is important to develop suitable support materials for catalysts in DMFC, which currently attracts a lot of attention.

A variety of support materials for PtRu catalysts such as Carbon Vulcan, CNTs, graphene, and oxides has been developed and used. **Table 3** and **Figure 5** show that the catalytic performance, durability, and even CO-tolerance varied depending on the support catalysts.

By comparing the different morphologies of carbon supports, Alegre and Kang's group found that methanol oxidation of carbon xerogels and carbon nanofiber substrates was respectively about 1.2 to 1.3 times higher than that of carbon substrate [74, 75]. Meanwhile, the CMK-8-II mesoporous carbon substrate fabricated by Maiyalagan's group gave superior methanol oxidation activity up to 2.2 times over carbon substrate [76]. CMK-8 mesoporous carbons with 3-D cubic Ia3d mesostructure consists of two interpenetrating continuous networks of chiral channels also called bicontinuous gyroidal. This structure possesses high surface area, well-defined pore size, high thermal stability, flexible framework composition, and intrinsic conductivity. With its distinct structure, CMK-8 mesoporous carbons have better diffusion and deposition of nanoparticles. In addition, the highly conductive Ia3d symmetric structure enhances electron transfer. Meanwhile, carbon xerogels have excellent properties such as high surface area, mesopore structure and high purity, allowing for high dispersion and efficient diffusion.

For the substrate of CNTs, the CV results show that the methanol oxidation capacity of the catalyst on CNTs substrate was 1.3–1.6 times higher than carbon Vulcan substrate [77–79]. CNTs possesses the higher graphite degree and the better electrical conductivity when it had the higher diffraction peaks than carbon black. Moreover, the smaller average particle size of PtRu/CNTs than that of pure Pt was one of the reasons for the high methanol oxidation capacity of this material system. In addition, a large specific surface area and high mechanical strength are also the preeminent properties of PtRu/CNTs that is attributed to the improved performance of the DMFC using PtRu/CNTs.

Graphene with its unique properties such as outstanding surface area, consistent porosity, good electrical conductivity and rich surface chemistry has become an excellent candidate for DMFC. Wang and Lee et al. reported that the current densities of PtRu/G or PtRu/GS were about 4 times and 1.5 times higher than those

| Catalyst | Support | Particle/crystal size (nm) | Catalyst performance | | Reference |
|---|---|---|---|---|---|
| | | | Measurement condition | Current density | |
| PtRu/E-Tek | Vulcan XC-72R | - | 2 M $CH_3OH$ + 0.5 M $H_2SO_4$, $\upsilon$ = 0.02 V $s^{-1}$ | 0.29 mA/$cm^2$ | [74] |
| PtRu/CX | Carbon xerogels | 3.5 | | 0.36 mA/$cm^2$ | |
| PtRu/C | Carbon | 2.0 | 1 M $CH_3OH$ and 0.5 M $H_2SO_4$, 2 mV sL1 | 340 mA/$cm^2$ | [75] |
| PtRu 70%/CNF | Carbon nanofiber | 2.9 | | 390 mA/$cm^2$ | |
| PtRu/XC-72 | Vulcan XC-72R Mesoporous Carbon | 4.0 | 0.5 M $H_2SO_4$ + 1 mol $dm^{-3}$ methanol at 50 $mV^{-1}$ | 27 mA/$cm^2$ | [76] |
| PtRu/CMK-8-II | | 5.0 | | 60 mA/$cm^2$ | |
| PtRu/C | Vulcan XC-72R | 3.0 | 20 mV/s in 1 M $CH_3OH$ + 0.5 M $H_2SO_4$ | 22.5 mA/$cm^2$ | [77] |
| PtRu/CNTs | Carbon nanotube | 2.5 | | 33.5 mA/$cm^2$ | |
| PtRu/CB | Carbon | 4.5 | 0.5 M $H_2SO_4$ and 1 M methanol, 50 mV/s | 27.5 mA $cm^{-2}$ | [78] |
| PtRu/CNT | Carbon nanotube | 3.9 | | 44.1 mA $cm^{-2}$ | |
| PtRu/N-CNTs | Carbon nanotube doping N | 3.5 | | 82.7 mA $cm^{-2}$ | |
| PtRu/C | Carbon | - | 1 M $CH_3OH$ + 0.5 M $H_2SO_4$, 20 mV $s^{-1}$ | 42.7 mA $mg^{-1}$ | [79] |
| PtRu/CNTs | Carbon nanotubes | - | | 56.0 mA $mg^{-1}$ | |
| PtRu/GS | Graphene sheet | 2.9 | | 78.7 mA $mg^{-1}$ | |
| PtRu/CTNs-GS | Carbon nanotubes + Graphene sheet | 2.4 | | 136.7 mA $mg^{-1}$ | |
| PtRu/MWCNTs | Carbon nanotubes | 3.75 | $N_2$-saturated, 0.5 M $H_2SO_4$ + 1.0 M $CH_3OH$, 10 mV/s | 4.82 mA/$cm^2$ | [80] |
| PtRu/Graphene | Graphene | 2.25 | | 20.8 mA/$cm^2$ | |
| PtRu/C | Carbon | - | 0.5 M $H_2SO_4$ + 1 M $CH_3OH$ | 430 mA $mg^{-1}$ | [81] |
| PtRu/RGO | Reduced graphene oxide (RGO) | 2.8 | | 570 mA $mg^{-1}$ | |
| PtRu/C | Carbon | 4.39 | 1 M $CH_3OH$ + 0.5 M $H_2SO_4$, 50 mV $s^{-1}$ | 8.21 mA $cm^{-2}$ | [82] |
| PtRu/FGSs | Functionalized graphene sheets | 2.87 | | 14.05 mA $cm^{-2}$ | |
| PtRu/AO-MWCNTs | Carbon nanotubes | 4.19 | $N_2$-saturated, 0.5 M $H_2SO_4$ + 1.0 M $CH_3OH$, 50 mV $s^{-1}$. | 112 mA/$cm^2$ | [83] |
| PtRu/PEI-MWCNTs | Functionalized carbon nanotubes | 3.17 | | 636 mA/$cm^2$ | |

**Table 3.**
*Results of methanol oxidation ability by CV measurement of PtRu on different supports.*

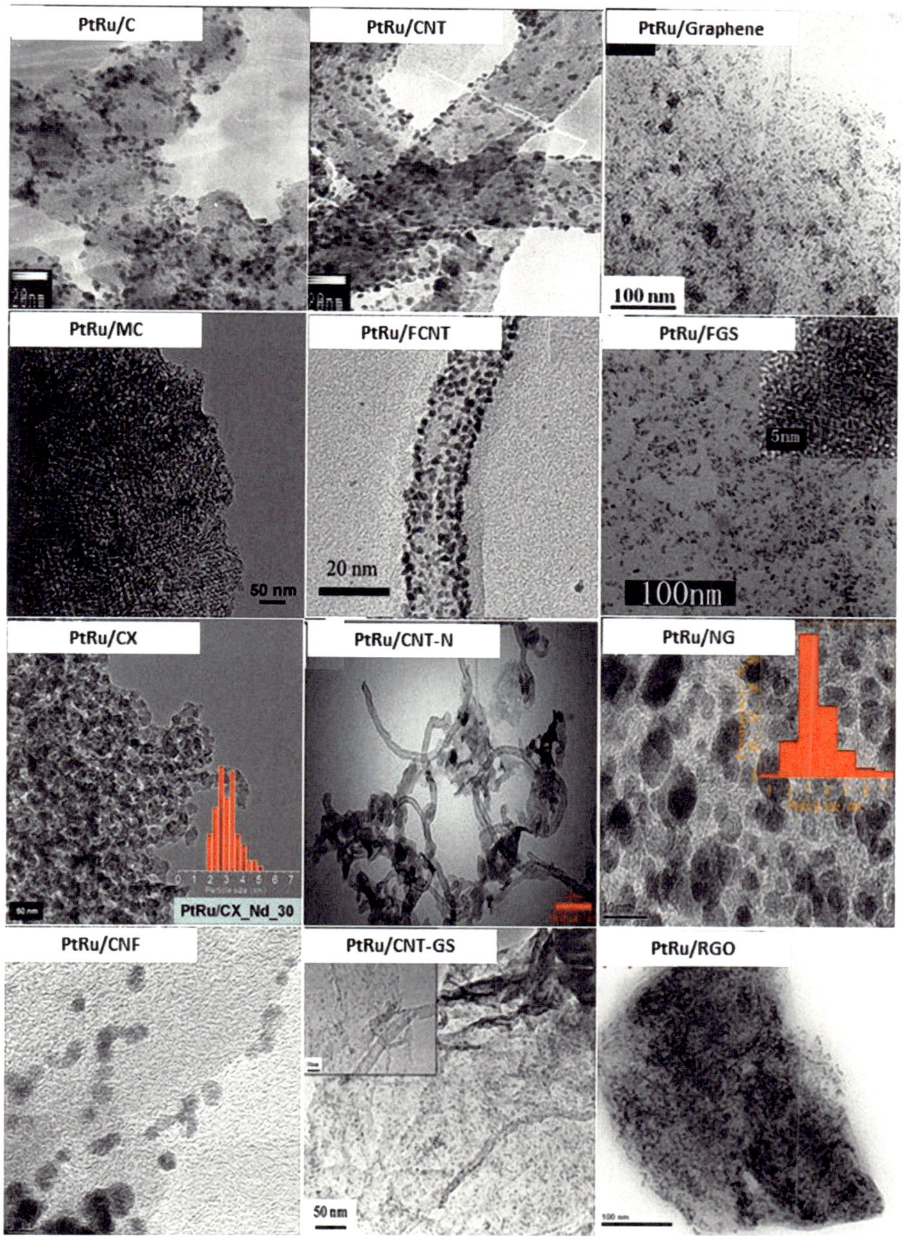


**Figure 5.**
*TEM images of PtRu on different supports: Carbon (C), mesoporous carbon (MC), carbon xerogels (CX), carbon nanofiber (CNF), carbon nanotube (CNT), functionalized carbon nanotubes (FCNT), carbon nanotube doping N (CNT-N), carbon nanotube - graphene sheet (CNT-GS), graphene, functionalized graphene sheet (FGS), graphene doping N (NG) and reduced graphene oxide (RGO) [74–83].*

of PtRu/CNTs [79, 80]. Similarly, the current density of reduced graphene oxide (RGO) substrate was about 1.3 times higher than carbon substrate one [81].

For comparison between carbon Vulcan substrate and graphene substrate, Zhao et al. found that the functionalized graphene sheets (FGSs) were 1.7 times higher in current density than carbon substrates [82]. The current density of N-doped carbon nanotube substrates was nearly 3 times higher than that of the conventional carbon substrates [78]. Due to the lack of bonding sites, namely -COOH, C = O and -OH groups on CNTs, the deposition of metal nanoparticles on the surface of CNTs is very difficult. To solve this problem, various methods have been implemented to functionalize CNTs in order to hold metal nanoparticles on its surfaces. Commonly, $HNO_3$ or $H_2SO_4$ was used to activate CNTs at an appropriate temperature and processing time. In addition, the functionalization of CNTs by various surfactants, aromatic compounds, functional polymers, and biomolecules have been proposed [83–86]. These methods can maintain the CNTs properties, allows to control the particle size, and create more uniform distribution for enhancing the DMFC efficiency. Consequently, the functionalized CNTs and graphene generally presented the improved methanol oxidation activity.

Combination of various materials is an approach toward developing effective support materials. For example, a highly porous architecture which was formed by combining CNTs and graphene sheets, provided a large exposure surface area, so the aggregation of metallic nanoparticles decreased to result in the enhanced catalytic performance [79]. Indeed, composite substrate of CNTs and graphene sheet resulted in current intensity 3.2 times higher than that of carbon substrate [79], suggesting that composite substrate materials can complement each component's advantages and suppress the disadvantages of each ingredient.

### 3.2 Effects of shape and structure on methanol oxidation of catalysts

Surface structure of catalysts is closely related to the catalytic activity. It is essential to find the optimal surface structures that can enhance methanol oxidation efficiency. For manipulating the morphology of nanoparticles, the key fabricating factors include precursors, ligand, capping agent, reductant, reaction time, and temperature. Up to now, nanoparticle catalysts have been successfully synthesized with many different morphologies such as nanopolyhedrons, nanowires, nanoplatelets, porous structure, hollow structure, concave structure, nanoframes, dendrites, monolayer, and core-shell structure [87–90]. **Figure 6** shows several various interesting morphologies of PtRu and their CV results [87–90]. Obviously, the maximum current densities $j_f$ of the new morphologies were higher than that of the traditional catalysts on carbon substrates. Because the new morphologies resulted in the differences in CO poisoning tolerance and methanol oxidation efficiency. Specifically, PtRu nanowires (NWs), PtRu nanorods (NRs) and PtRu nanocubes (NCs) were successfully fabricated by a one-step solvothermal method [87]. As shown in **Figure 6**, the PtRu NWs possessed higher methanol oxidation reaction activity than PtRu NRs and PtRu NCs. In addition, the onset potential of PtRu NWs is lower and its $j_f/j_b$ ratio is greater compared with PtRu NRs and PtRu NCs. Based on density functional theory, calculations show a transition state (Pt-CO ••• OH-Ru) that is only formed from the interaction between Pt-COads and Ru-OHads species if the distance between Pt and Ru atoms is less than or equal to 4 Å. Therefore, the close connection between Pt and Ru atoms in PtRu alloy structure is more beneficial than heterostructure. In addition, the methanol oxidation catalytic activity for PtRu NWs was higher than that of PtRu NCs because the adsorption energy of $CO_{ads}$ and $OH_{ads}$ on {111} facets enclosed PtRu NWs reached near the optimal value compared to the adsorption energy of $CO_{ads}$ and $OH_{ads}$ on {100} facets enclosed PtRu NCs. Moreover, compared with PtRu NRs, the PtRu NWs had more {111} active sites facets due to their longer and thinner structure to result in the higher electrocatalytic activity. **Figure 6** also shows that PtRu nanodendrites have higher methanol oxidation efficiency than PtRu NCs or Pt/C because PtRu nanodendrites with the staggered branches can facilitate guest accessibility and tolerance to undesired agglomeration [89]. Furthermore, PtRu nanodendrites has surface area and atom utility is larger than PtRu NCs to get

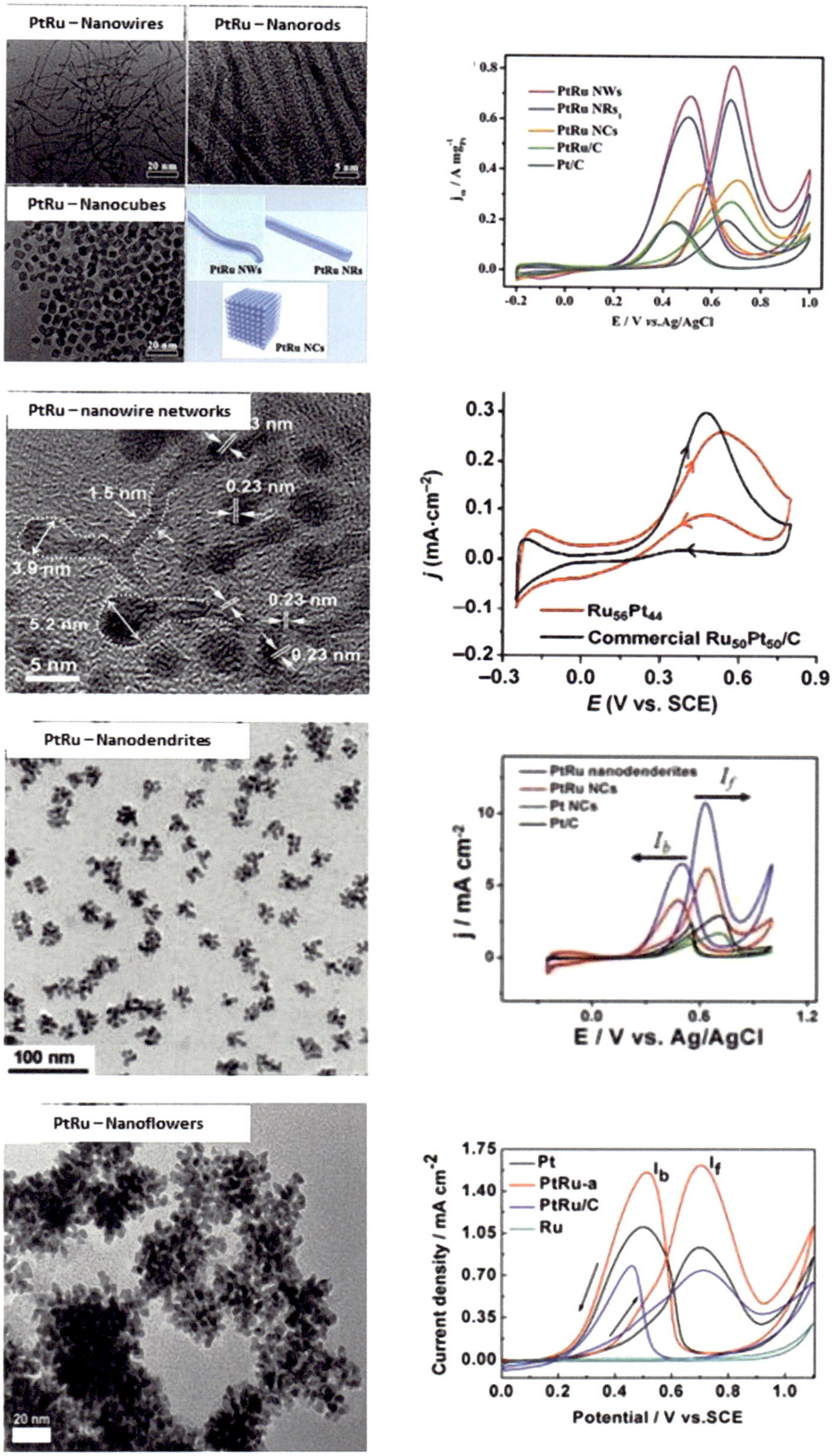


**Figure 6.**
*TEM images and CV curves of the various morphologies of PtRu [87–90].*

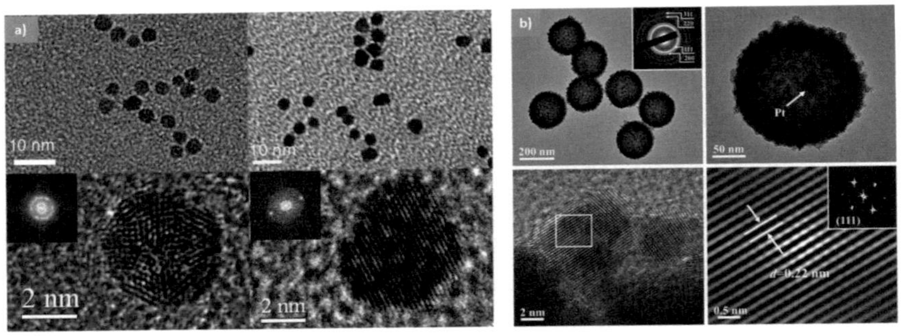

**Figure 7.**
*TEM and HR-TEM images of core-shell (a) and Pt@mPtRu Yolk – Shell (b) nanostructures [93, 94].*

better facilitate electron catalysis. PtRu nanoflower catalysts synthesized by a facile one-pot solvothermal method exhibited superior methanol oxidation efficiency compared to PtRu/C, as shown in **Figure 6** [90].

In order to improve fuel cell performance and reduce their cost, the fuel cell catalysis primarily needs to optimize its composition and structure – morphology. The core-shell structure has attracted much attention owing to its wide applications in electrochemical energy devices. Various works on Ru@Pt nanoparticles have demonstrated that Ru@Pt has superior CO tolerance and higher methanol oxidation efficiency than that of PtRu [91–94]. A demonstrated Ru@Pt structure is shown in **Figure 7**. An increase of active sites is a necessary requirement to enhance the Pt-based electrocatalyst utilization. Hollow nanostructures not only meet this requirement but also reduce the amount of Pt, while porous nanostructures can reduce diffusion resistance and increase contact area [92, 93]. A combination of core-shell and hollow structures is of great interest and become a potential new research direction. Recently, yolk–shell nanostructures have been successfully synthesized by a facile approach (**Figure 7**) [94]. As a result, this structure has better catalytic activity, durability and tolerance to CO comparing to PtRu nanocages and Pt/C [94].

## 4. Conclusion

In summary, the two-component Pt alloy has generally better methanol oxidation efficiency and tolerances to CO poisoning than pure Pt. Among PtM alloys, PtRu is the most notable due to its superior properties of a low potential onset, high catalytic activity, and high CO tolerance. The performance of DMFC can be further improved when PtRu is combined with a third suitable metal. Up to now, PtRuM nanoparticles with different morphologies have been successfully synthesized, and they exhibited higher methanol oxidation activities than those of traditional PtRu. Substrate and morphology are also the important factors in manipulating fuel cell performance, and thus a variety of substrates has been used for DMFC catalyst. It is found that CNTs and graphene are the two most common support materials to mitigate the disadvantages of carbon Vulcan (a traditional support material (or substrate)) and thereby enhancing the fuel cell efficiency. Functionalization and doping approaches are the next developments to improve and search for the optimal substrates. Interestingly, the combination of CNTs and graphene substrates possesses as an excellent substrate for methanol oxidation. The demonstrated results in this book chapter open up a new research direction that involves in simultaneously optimization of nanocatalysts, substrates, and their structures – morphologies toward the developments of DMFC.

## Acknowledgements

This research is funded by Vietnam National Foundation for Science and Technology Development (NAFOSTED) under grant number 103.02-2019.374, and Ho Chi Minh City University of Technology (HCMUT), VNU-HCM under grant number BK-SDH-2021-2080906. We acknowledge the support of time and facilities from Ho Chi Minh City University of Technology (HCMUT), VNU-HCM for this study.

## Author details

Dang Long Quan[1,2,3] and Phuoc Huu Le[4*]

1 Ho Chi Minh City University of Technology (HCMUT), Ho Chi Minh City, Vietnam

2 Vietnam National University Ho Chi Minh City, Ho Chi Minh City, Vietnam

3 Department of Physics, College of Natural Sciences, Can Tho University, Can Tho City, Vietnam

4 Department of Physics and Biophysics, Faculty of Basic Sciences, Can Tho University of Medicine and Pharmacy, Can Tho, Vietnam

*Address all correspondence to: lhuuphuoc@ctump.edu.vn

## References

[1] Almeida TS, Garbim C, Silva RG, De Andrade AR. Addition of iron oxide to Pt-based catalyst to enhance the catalytic activity of ethanol electrooxidation. Journal of Electroanalytical Chemistry. 2017; *796*: 49-56.

[2] Zhang C, Shen X, Pan Y, Peng Z. A review of Pt-based electrocatalysts for oxygen reduction reaction. Frontiers in Energy. 2017; *11*: 268-285.

[3] Seselj N, Engelbrekt C, Zhang, J. Graphene-supported platinum catalysts for fuel cells. Science Bulletin. 2015; *60*: 864-876.

[4] Aricò AS, Baglio V, Antonucci V. Direct methanol fuel cells: history, status and perspectives. Electrocatalysis of direct methanol fuel cells. Wiley 2009; 1-78. ISBN: 9783527627707

[5] Tiwari JN, Tiwari RN, Singh G, Kim KS. Recent progress in the development of anode and cathode catalysts for direct methanol fuel cells. Nano Energy. 2013; *2*: 553-578.

[6] Ren X, Lv Q, Liu L, Liu B, Wang Y, Liu A, Wu G. Current progress of Pt and Pt-based electrocatalysts used for fuel cells. Sustainable Energy & Fuels. 2020; 4: 15-30.

[7] Janssen MMP, Moolhuysen J. Platinum—tin catalysts for methanol fuel cells prepared by a novel immersion technique, by electrocodeposition and by alloying. Electrochimica Acta. 1976; 21: 861-868.

[8] Watanabe M, Motoo S. Electrocatalysis by ad-atoms: Part III. Enhancement of the oxidation of carbon monoxide on platinum by ruthenium ad-atoms. J. Electroanal. Chem. Interf. Electrochem. 1975; *60*: 275-283.

[9] Nassr ABAA, Sinev I, Grünert W, Bron M. PtNi supported on oxygen functionalized carbon nanotubes: in depth structural characterization and activity for methanol electrooxidation. Appl Catal B Environ. 2013; *142-143*:849-860.

[10] Baronia R, Goel J, Tiwari S, Singh P, Singh D, Singh SP, Singhal SK. Efficient electro-oxidation of methanol using PtCo nanocatalysts supported reduced graphene oxide matrix as anode for DMFC. Int. J. Hydrog Energy. 2017; *42*:10238-10247.

[11] Baglio V, Stassi A, Di Blasi A., Durso C, Antonucci V, Aricò AS. Investigation of bimetallic Pt–M/C as DMFC cathode catalysts. Electrochimica Acta. 2007; 53(3):1360-1364.

[12] Sahin O, Kivrak H.. A comparative study of electrochemical methods on Pt–Ru DMFC anode catalysts: the effect of Ru addition. International journal of hydrogen energy. 2013; 38(2):901-909.

[13] Jia H, Chang G, Shu H, Xu M, Wang X, Zhang Z, Liu X, He H, Wang K, Zhu R, He Y. Pt nanoparticles modified Au dendritic nanostructures: facile synthesis and enhanced electrocatalytic performance for methanol oxidation. Int. J. Hydrog Energy. 2017; 42: 22100-22107.

[14] Alekseenko AA, Guterman V.E., Belenov S.V., Menshikov V. S., Tabachkova N.Y., Safronenko O.I., Moguchikh E.A. Pt/C electrocatalysts based on the nanoparticles with the gradient structure. International Journal of Hydrogen Energy. 2018; 43(7): 3676-3687.

[15] Xu J, Hua K, Sun G, Wang C, Lv X, Wang Y. Electrooxidation of methanol on carbon nanotubes supported Pt–Fe alloy electrode. Electrochemistry communications. 2006; 8(6): 982-986.

[16] Baronia R, Goel J, Tiwari S, Singh P, Singh D, Singh SP, Singhal SK.

Efficient electro-oxidation of methanol using PtCo nanocatalysts supported reduced graphene oxide matrix as anode for DMFC. International Journal of Hydrogen Energy. 2017; 42(15): 10238-10247.

[17] Nassr A.B.A.A, Sinev I, Grünert W, Bron M. PtNi supported on oxygen functionalized carbon nanotubes: in depth structural characterization and activity for methanol electrooxidation. Applied Catalysis B: Environmental. 2013; 142: 849-860.

[18] Li F, Guo Y, Chen M, Qiu H, Sun X, Wang W, Gao J. Comparison study of electrocatalytic activity of reduced graphene oxide supported Pt–Cu bimetallic or Pt nanoparticles for the electrooxidation of methanol and ethanol. International journal of hydrogen energy. 2013; 38(33): 14242-14249.

[19] Maya-Cornejo J, Garcia-Bernabé A, Compañ V. Bimetallic Pt-M electrocatalysts supported on single-wall carbon nanotubes for hydrogen and methanol electrooxidation in fuel cells applications. International Journal of Hydrogen Energy. 2018; 43(2): 872-884.

[20] Liu Z, Ling XY, Guo B, Hong L, Lee JY. Pt and PtRu nanoparticles deposited on single-wall carbon nanotubes for methanol electro-oxidation. Journal of power sources. 2007; 167(2): 272-280.

[21] Lu Q, Huang J, Han C, Sun L, Yang X. Facile synthesis of composition-tunable PtRh nanosponges for methanol oxidation reaction. Electrochimica Acta 2018; 266: 305-311.

[22] Zhao H, Qi W, Zhou X, Wu H, Li Y. Composition-controlled synthesis of platinum and palladium nanoalloys as highly active electrocatalysts for methanol oxidation. Chinese Journal of Catalysis. 2018; 39(2): 342-349.

[23] Wroblowa H. Electrocatalysis. Journal of Electroanalytical Chemistry. 1964; 7(6): 428-451.

[24] Petrii OA. Pt–Ru electrocatalysts for fuel cells: a representative review. Journal of Solid State Electrochemistry. 2008; 12(5): 609.

[25] Nørskov JK, Rossmeisl J, Logadottir A, Lindqvist LRKJ, Kitchin JR, Bligaard T, Jonsson H. Origin of the overpotential for oxygen reduction at a fuel-cell cathode. The Journal of Physical Chemistry B 2004; 108(46): 17886-17892.

[26] Watanabe M, MoToo S. Correlation between Electro-Catalytic Activity of Platinum-Ruthenium Alloys and Coverage of Organic Substances. Denki Kagaku oyobi Kogyo Butsuri Kagaku. 1973; 41(3): 190-196.

[27] Watanabe MA, Motoo S. Electrocatalysis by ad-atoms: Part II. Enhancement of the oxidation of methanol on platinum by ruthenium ad-atoms. J. Electroanal. Chem. Interf. Electrochem. 1975; 60(3): 267-273.

[28] Ramli ZAC, Kamarudin SK. Platinum-based catalysts on various carbon supports and conducting polymers for direct methanol fuel cell applications: a review. Nanoscale research letters. 2018; *13*(1): 410.

[29] Stamenkovic V, Mun BS, Mayrhofer KJ, Ross PN, Markovic NM, Rossmeisl J, Nørskov JK. Changing the activity of electrocatalysts for oxygen reduction by tuning the surface electronic structure. Angewandte Chemie. 2006; 118: 2963-2967.

[30] Strasser P, Fan Q, Devenney M, Weinberg WH, Liu P, Nørskov JK. High throughput experimental and theoretical predictive screening of materials– a comparative study of search strategies for new fuel cell anode catalysts. The Journal of Physical Chemistry B. 2003; 107: 11013-11021.

[31] Samjeské G, Wang H, Löffler T, Baltruschat H. CO and methanol oxidation at Pt-electrodes modified by Mo. Electrochimica Acta. 2002; *47*: 3681-3692.

[32] Ordóñez LC, Roquero P, Sebastian PJ, Ramírez J. CO oxidation on carbon-supported PtMo electrocatalysts: effect of the platinum particle size. International journal of hydrogen energy. 2007; 32: 3147-3153.

[33] Liu Z, Hu JE, Wang Q, Gaskell K, Frenkel AI, Jackson GS, Eichhorn B. PtMo alloy and MoOx@ Pt Core– shell nanoparticles as highly CO-tolerant electrocatalysts. Journal of the American Chemical Society 2009; 131: 6924-6925.

[34] Zhou ZH, Li WS, Fu Z, Xiang XD. Carbon nanotube-supported Pt-$HxMoO_3$ as electrocatalyst for methanol oxidation. International Journal of Hydrogen Energy. 2010; *35*(3): 936-941.

[35] Ordóñez LC, Roquero P, Sebastian PJ, Ramírez J. Carbon-supported platinum-molybdenum electro-catalysts for methanol oxidation. Catalysis Today. 2005; *107*: 46-52.

[36] Mukerjee S, Urian RC. Bifunctionality in Pt alloy nanocluster electrocatalysts for enhanced methanol oxidation and CO tolerance in PEM fuel cells: electrochemical and in situ synchrotron spectroscopy. Electrochimica Acta. 2002; *47*(19): 3219-3231.

[37] Tsiouvaras N, Martínez-Huerta MV, Paschos O, Stimming U, Fierro JLG, Peña MA. PtRuMo/C catalysts for direct methanol fuel cells: Effect of the pretreatment on the structural characteristics and methanol electrooxidation. International journal of hydrogen energy. 2010; *35*(20):11478-11488.

[38] Chen S, Ye F, Lin W. Effect of operating conditions on the performance of a direct methanol fuel cell with PtRuMo/CNTs as anode catalyst. International Journal of Hydrogen Energy. 2010; *35*(15): 8225-8233.

[39] Kakati N, Maiti J, Oh JY, Yoon YS. Study of methanol oxidation of hydrothermally synthesized PtRuMo on multi wall carbon nanotubes. Applied Surface Science. 2011; *257*: 8433-8437.

[40] Park SH, Jung HM, Um S, Song YW, Kim HS. Rapid synthesis of Pt-based alloy/carbon nanotube catalysts for a direct methanol fuel cell using flash light irradiation. International Journal of Hydrogen energy. 2012; *37*(17): 12597-12604.

[41] Watanabe M, Igarashi H, Fujino T. Design of CO tolerant anode catalysts for polymer electrolyte fuel cell. Electrochemistry. 1999; *67*(12): 1194-1196.

[42] Watanabe M, Zhu Y, Uchida H. Oxidation of CO on a Pt– Fe alloy electrode studied by surface enhanced infrared reflection– absorption spectroscopy. The Journal of Physical Chemistry B. 2000; *104*(8): 1762-1768.

[43] Jeon MK, Won JY, Lee KR, Woo SI. Highly active PtRuFe/C catalyst for methanol electro-oxidation. Electrochemistry Communications. 2007; *9*(9): 2163-2166.

[44] Cai Z, Kuang Y, Qi X, Wang P, Zhang Y, Zhang Z, Sun X. Ultrathin branched PtFe and PtRuFe nanodendrites with enhanced electrocatalytic activity. Journal of Materials Chemistry A. 2015; 3(3): 1182-1187.

[45] Huang T, Wang X, Zhuang J, Cai WB, Yu A. Preparation of Porous PtRuCo Catalyst by One-Step

Codeposition and Its Electrocatalytic Performance for Methanol Oxidation. Electrochemical and Solid State Letters. 2009; *12*(7): B112.

[46] Ahn SH, Kwon OJ, Kim SK, Choi I, Kim JJ. Electrochemical preparation of Pt-based ternary alloy catalyst for direct methanol fuel cell anode. International Journal of Hydrogen Energy 2010; *35*(24): 13309-13316.

[47] Paulus UA, Wokaun A, Scherer GG, Schmidt TJ, Stamenkovic V, Radmilovic V, Markovic NM, Ross PN. Oxygen reduction on carbon-supported Pt–Ni and Pt–Co alloy catalysts. The Journal of Physical Chemistry B. 2002; 106: 4181-4191.

[48] Park KW, Choi JH, Kwon BK, Lee SA, Sung YE, Ha HY, Hong SA, Kim H, Wieckowski A. Chemical and electronic effects of Ni in Pt/Ni and Pt/Ru/Ni alloy nanoparticles in methanol electrooxidation. The Journal of Physical Chemistry B. 2002; 106: 1869-1877.

[49] Wang ZB, Yin GP, Shi PF, Sun YC. Novel Pt–Ru–Ni/C catalysts for methanol electro-oxidation in acid medium. Electrochemical and Solid State Letters. 2005; *9*(1): A13.

[50] Wang ZB, Yin GP, Zhang J, Sun YC, Shi PF. Co-catalytic effect of Ni in the methanol electro-oxidation on Pt–Ru/C catalyst for direct methanol fuel cell. Electrochimica Acta. 2006; *51*(26): 5691-5697.

[51] Ye F, Chen S, Dong X, Lin W. Carbon nanotubes supported Pt-Ru-Ni as methanol electro-oxidation catalyst for direct methanol fuel cells. Journal of Natural Gas Chemistry. 2007; *16*(2): 162-166.

[52] Liang Y, Zhang H, Tian Z, Zhu X, Wang X, Yi B.. Synthesis and structure– activity relationship exploration of carbon-supported PtRuNi nanocomposite as a CO-tolerant electrocatalyst for proton exchange membrane fuel cells. The Journal of Physical Chemistry B. 2006; 110: 7828-7834.

[53] Wang ZB, Yin GP, Shao YY, Yang BQ, Shi PF, Feng PX. Electrochemical impedance studies on carbon supported PtRuNi and PtRu anode catalysts in acid medium for direct methanol fuel cell. Journal of Power Sources. 2007; *165*(1): 9-15.

[54] Ye F, Chen S, Dong X, Lin W. Carbon nanotubes supported Pt-Ru-Ni as methanol electro-oxidation catalyst for direct methanol fuel cells. Journal of Natural Gas Chemistry. 2007; 16(2): 162-166.

[55] Roth C, Goetz M, Fuess H. (2001). Synthesis and characterization of carbon-supported Pt–Ru–WOx catalysts by spectroscopic and diffraction methods. Journal of Applied Electrochemistry; 31(7): 793-798.

[56] Xue S, Deng W, Yang F, Yang J, Amiinu IS, He D, Tang H, Mu S. Hexapod PtRuCu nanocrystalline alloy for highly efficient and stable methanol oxidation. ACS Catalysis. 2018; 8(8): 7578-7584.

[57] Jeon MK, Lee KR, Woo SI. Ternary Pt 45 Ru 45 M 10/C (M = Mn, Mo and W) catalysts for methanol and ethanol electro-oxidation. Korean journal of chemical engineering. 2009; 26(4): 1028-1033.

[58] Yu X, Ye S. Recent advances in activity and durability enhancement of Pt/C catalytic cathode in PEMFC: Part I. Physico-chemical and electronic interaction between Pt and carbon support, and activity enhancement of Pt/C catalyst. Journal of power sources. 2007; 172(1): 133-144.

[59] Jung N, Chung DY, Ryu J, Yoo SJ, Sung YE. Pt-based nanoarchitecture and catalyst design for fuel cell applications. Nano Today. 2014; 9(4): 433-456.

[60] Antolini E. Carbon supports for low-temperature fuel cell catalysts. Appl. Catal. B. 2009; 88(1-2): 1-24.

[61] Yang WH, Zhang QH, Wang HH, Zhou ZY, Sun SG. Preparation and utilization of a sub-5 nm $PbO_2$ colloid as an excellent co-catalyst for Pt-based catalysts toward ethanol electro-oxidation. New Journal of Chemistry. 2017; 41(20): 12123-12130.

[62] Antoniassi RM, Silva JCM, Neto AO, Spinacé EV. Synthesis of Pt+$SnO_2$/C electrocatalysts containing Pt nanoparticles with preferential (100) orientation for direct ethanol fuel cell. Applied Catalysis B: Environmental. 2017; 218: 91-100.

[63] Luo Y, Alonso-Vante N. The effect of support on advanced Pt-based cathodes towards the oxygen reduction reaction. State of the art. Electrochimica Acta. 2015; 179: 108-118.

[64] Shao Y, Yin G, Gao Y, Shi P. Durability study of Pt/ C and Pt/ CNTs catalysts under simulated PEM fuel cell conditions. Journal of the Electrochemical Society. 2006; 153(6): A1093.

[65] Li W, Liang C, Zhou W, Qiu J, Zhou Z, Sun G, Xin Q. Preparation and characterization of multiwalled carbon nanotube-supported platinum for cathode catalysts ofdirect methanol fuel cells. J. Phys. Chem. B. 2003; 107: 6292-6299.

[66] Tian ZQ, Jiang SP, Liang YM, Shen PK. Synthesis and characterization of platinum catalysts on multiwalled carbon nanotubes by intermittent microwave irradiation for fuel cell applications. J. Phys. Chem. B. 2006; 110: 5343-5350.

[67] Li W, Liang C, Qiu J, Zhou W, Han H, Wei Z, Sun G, Xin Q. Carbon nanotubes as support for cathode catalyst of a direct methanol fuel cell. Carbon. 2002; 40: 791-794.

[68] Li L, Xing Y. Electrochemical durability of carbon nanotubes at 80°C. J. Power Sources. 2008; 178: 75-79.

[69] Li L, Xing Y. Electrochemical durability of carbon nanotubes in noncatalyzed and catalyzed oxidations. J. Electrochem. Soc. 2006; 153: A1823–A1828

[70] Kongkanand A, Kuwabata S, Girishkumar G, Kamat P. Single-wall carbon nanotubes supported platinum nanoparticles with improved electrocatalytic activity for oxygen reduction reaction. Langmuir. 2006; 22: 2392-2396.

[71] Shao Y, Yin G, Gao Y, Shi P. Durability study of Pt/C and Pt/CNTs catalysts under simulated PEM fuel cell conditions. J. Electrochem. Soc. 2006; 153: A1093–A1097.

[72] Wang X, Li W, Chen Z, Waje M, Yan Y. Durability investigation of carbon nanotube as catalyst support for proton exchange membrane fuel cell. J. Power Sources. 2006; 158: 154-159.

[73] Rambabu G, Bhat DS, Figueiredo FM. Carbon Nanocomposite Membrane Electrolytes for Direct Methanol Fuel Cells—A Concise Review. Nanomaterials. 2019; 9(9): 1292.

[74] Alegre C, Calvillo L, Moliner R, González-Expósito JA, Guillén-Villafuerte O, Huerta MM, Pastor E, Lázaro MJ. Pt and PtRu electrocatalysts supported on carbon xerogels for direct methanol fuel cells. Journal of Power Sources. 2011; *196*(9): 4226-4235.

[75] Kang S, Lim S, Peck DH, Kim SK, Jung DH, Hong SH, Jung HG, Shul Y.

Stability and durability of PtRu catalysts supported on carbon nanofibers for direct methanol fuel cells. International journal of hydrogen energy. 2012; *37*(5): 4685-4693.

[76] Maiyalagan T, Alaje TO, Scott K. Highly stable Pt–Ru nanoparticles supported on three-dimensional cubic ordered mesoporous carbon (Pt–Ru/CMK-8) as promising electrocatalysts for methanol oxidation. The Journal of Physical Chemistry C. 2012; 116(3): 2630-2638.

[77] Cui Z, Liu C, Liao J, Xing W. Highly active PtRu catalysts supported on carbon nanotubes prepared by modified impregnation method for methanol electro-oxidation. Electrochimica Acta. 2008; *53*(27): 7807-7811.

[78] Fard HF, Khodaverdi M, Pourfayaz F, Ahmadi MH. Application of N-doped carbon nanotube-supported Pt-Ru as electrocatalyst layer in passive direct methanol fuel cell. International Journal of Hydrogen Energy. 2020; 45(46): 25307-25316.

[79] Wang YS, Yang SY, Li SM, Tien HW, Hsiao ST, Liao WH, Liu CH, Chang KH, Ma CCM, Hu CC. Three-dimensionally porous graphene–carbon nanotube composite-supported PtRu catalysts with an ultrahigh electrocatalytic activity for methanol oxidation. Electrochimica Acta. 2013; 87: 261-269.

[80] Lee SH, Kakati N, Jee SH, Maiti J, Yoon YS. Hydrothermal synthesis of PtRu nanoparticles supported on graphene sheets for methanol oxidation in direct methanol fuel cell. Materials Letters. 2011; 65: 3281-3284.

[81] Reddy GV, Raghavendra P, Ankamwar B, Chandana PS, Kumar SS, Sarma LS. Ultrafine Pt–Ru bimetallic nanoparticles anchored on reduced graphene oxide sheets as highly active electrocatalysts for methanol oxidation. Materials Chemistry Frontiers. 2017; 1(4): 757-766.

[82] Zhao J, Zhang L, Xue H, Wang Z, Hu H. Methanol electrocatalytic oxidation on highly dispersed platinum–ruthenium/graphene catalysts prepared in supercritical carbon dioxide–methanol solution. *RSC Advances*. 2012; 2(25): 9651-9659.

[83] Cheng Y. Highly effective and CO-tolerant PtRu electrocatalysts supported on poly(ethyleneimine) functionalized carbon nanotubes for direct methanol fuel cells. Electrochimica Acta. 2013; 99: 124-132.

[84] Prabhuram J, Zhao TS, Liang ZX, Chen R. A simple method for the synthesis of PtRu nanoparticles on the multi-walled carbon nanotube for the anode of a DMFC. Electrochimica Acta. 2007; *52*(7): 2649-2656.

[85] Carmo M, Paganin VA, Rosolen JM, Gonzalez ER. Alternative supports for the preparation of catalysts for low-temperature fuel cells: the use of carbon nanotubes. Journal of Power Sources. 2005; *142*(1-2): 169-176.

[86] Gan L, Lv R, Du H, Li B, Kang F. High loading of Pt–Ru nanocatalysts by pentagon defects introduced in a bamboo-shaped carbon nanotube support for high performance anode of direct methanol fuel cells. Electrochemistry Communications. 2009; *11*(2): 355-358.

[87] Huang L, Zhang X, Wang Q, Han Y, Fang Y, Dong S. Shape-control of Pt–Ru nanocrystals: tuning surface structure for enhanced electrocatalytic methanol oxidation. Journal of the American Chemical Society. 2018; 140(3): 1142-1147.

[88] Zhao W, Huang D, Yuan Q, Wang X. Sub-2.0-nm Ru and composition-tunable RuPt nanowire

networks. Nano Research. 2016; 9(10): 3066-3074.

[89] Lu S, Eid K, Ge D, Guo J, Wang L, Wang H, Gu H. One-pot synthesis of PtRu nanodendrites as efficient catalysts for methanol oxidation reaction. Nanoscale. 2017; 9(3): 1033-1039.

[90] Li M, Zheng H, Han G, Xiao Y, Li Y. Facile synthesis of binary PtRu nanoflowers for advanced electrocatalysts toward methanol oxidation. Catalysis Communications. 2017; 92: 95-99.

[91] Dang LQ, Nguyen MT, Truong NV, Le PH, Long N. Investigation of Carbon Supported Ru–Pt Nanoparticles for High–Performance Electrocatalytic Oxidation of Methanol. Int. J. Electrochem. Sci. 2017; 12: 10187-10198.

[92] Yin HJ, Zhou JH, Zhang YW. Shaping well-defined noble-metal-based nanostructures for fabricating high-performance electrocatalysts: advances and perspectives. Inorganic Chemistry Frontiers. 2019; 6(10): 2582-2618.

[93] Alayoglu S, Zavalij P, Eichhorn B, Wang Q, Frenkel AI, Chupas P. Structural and architectural evaluation of bimetallic nanoparticles: a case study of Pt– Ru core– shell and alloy nanoparticles. Acs Nano. 2009; 3(10): 3127-3137.

[94] Yin S, Kumar RD, Yu H, Li C, Wang Z, Xu Y, Li X, Wang L, Wang H. Pt@Mesoporous PtRu Yolk–Shell Nanostructured Electrocatalyst for Methanol Oxidation Reaction. ACS Sustainable Chemistry & Engineering. 2019; 7(17): 14867-14873.

Chapter 2

# Nickel Foam Electrode with Low Catalyst Loading and High Performance for Alkaline Direct Alcohol Fuel Cells

*Qian Xu, Jiajia Zhang and Chunzhen Yang*

## Abstract

Nickel foam has a unique three-dimensional (3-D) network structure that helps to effectively utilize catalysts and is often used as an electrode support material for alkaline direct alcohol fuel cells. In this chapter, first, the effect of nickel foam thickness on cell performance is explored. The results show that the thickness affects both mass transfer and electron conduction, and there is an optimal thickness. The thinner the nickel foam is, the better the conductivity is. However, the corresponding three-dimensional space becomes narrower, which results in a partial agglomeration of the catalyst and the hindrance of mass transfer. The cell performance of 0.6 mm nickel foam electrode is better than that of 0.3 and 1.0 mm. Secondly, to fully exert the catalytic function of the catalyst even at a lower loading, a mixed acid-etched nickel foam electrode with lower Pd loading (0.35 mg $cm^{-2}$) is prepared then by a spontaneous deposition method. The maximum power density of the single alkaline direct ethanol fuel cell (ADEFC) can reach 30 mW $cm^{-2}$, which is twice the performance of the hydrochloric acid treated nickel foam electrode. The performance improvement is attributed to the micro-holes produced by mixed acids etching, which enhances the roughness of the skeleton and improves the catalyst electrochemical active surface area.

**Keywords:** nickel foam, thickness, mixed acid, porous skeleton, direct alcohol fuel cell

## 1. Introduction

Due to the increasing energy consumption and the pollution by fossil fuels, the development of green and environmentally friendly renewable energy sources has won extensive supports from the international community. Ethanol is a kind of low toxicity, high-energy-density liquid fuel, which has the advantages of convenient storage, transportation, and portability. Moreover, ethanol has a wide range of sources, which can be produced from biomass such as corn and sugarcane. Alkaline direct ethanol fuel cell (ADEFC) is an electrochemical power generation device that directly uses ethanol and oxygen with oxidation and reduction reactions at the anode and cathode, respectively. It is an ideal portable power source in the future society with green and efficient characteristics [1–3].

The critical part of a fuel cell is the membrane electrode assembly (MEA), which includes a multi-layer structure of an anode diffusion layer, an anode catalyst layer, a polymer electrolyte membrane, a cathode catalyst layer, and a cathode diffusion layer. Among them, the diffusion layer holds the functions of transporting reactants and products, conducting electrons, and supporting the catalyst layer. Therefore, the performance of the diffusion layer is related to the performance of the fuel cell [4, 5].

The commonly used electrode support materials include carbon paper, carbon cloth, metal foam, etc. Carbon paper is made by carbonizing carbon fiber and binder together. Carbon paper has good uniformity. However, it has poor strength and is easy to break. The ribs of the flow channel on the bipolar plate are also easy to crush the carbon paper during the battery stack assembly process. Carbon cloth is woven from carbon fiber yarn, or is woven into cloth from carbon fiber precursors and then carbonized. Carbon cloth has good strength and is not easy to break. However, due to weaving, the surface flatness of carbon cloth is worse than that of carbon paper. Although the performance of carbon paper and carbon cloth is excellent, the production of carbon paper and carbon cloth requires high-temperature graphitization, the process is complicated, and the technical requirements are high, resulting in high prices of carbon paper and carbon cloth and increasing the cost of fuel cells [6, 7].

Metal foam is a porous metal material that forms countless three-dimensional (3-D) spatial network structures in a metal matrix. It is composed of a rigid framework and internal pores, because it has both metal characteristics and some special physical properties of non-metal, such as porous, lightweight, low density, and high specific strength. There are different types of metal foam, including aluminum foam, titanium foam, copper foam, iron foam, and nickel foam [8]. Among them, nickel foam (Ni foam) is often used as an electrode substrate material in electrochemical energy storage devices, such as secondary nickel-hydrogen, nickel-cadmium batteries, and nickel-zinc batteries, which play a role in collecting current and supporting active materials. The production of Ni foam mainly adopts the electrodeposition method, which uses polyurethane foam as the skeleton. After conductive treatment, electrodeposition and heat treatment are carried out to finally obtain Ni foam [9].

In recent years, Ni foam has been widely used in ADEFC as the electrode support material. The excellent physical properties such as three-dimensional structure, high porosity, high conductivity, and resistance in alkaline media, make the researches of Ni foam develop rapidly. Coupled with the low price of Ni foam, it is more conducive to commercialization.

So far, several works have reported on Ni foams used in fuel cell applications. In 2010, Wang et al. deposited palladium catalyst directly on Ni foam by electrodeposition and found that the three-dimensional porous electrode greatly improved the catalytic activity and stability for ethanol oxidation [10]. In 2011, Li et al. used Ni foam integrated electrode prepared by dip-coating method to perform better on ADEFC than conventional carbon paper electrode [11]. In 2019, Sun et al. further improved the cell performance by improving the electrode preparation method, and the power density reached 202 mW $cm^{-2}$ [12]. It can be found that previous work has demonstrated the application potential of porous electrodes.

By literature survey, it is found that the influence of the physical properties (thickness) of the Ni foam on the cell performance is still unknown. However, the thickness determines the size of the entire three-dimensional space, its own conductivity, and the length of the material transmission path. The thinner the Ni foam, the better the conductivity, but the corresponding three-dimensional space becomes narrower, which may cause agglomeration of part of the catalyst and hinder mass transfer.

At the same time, the smooth skeleton surface of pristine Ni foam results in a low specific surface area, such that a high-load catalyst is required to deal with ethanol oxidation, which increases the cost and limits its application as catalyst support [13]. Hence, efforts to enhance the roughness of the skeleton surface and reduce the catalyst loading have been intensively made. Since an inert layer (oxide layer or hydroxide layer) is easily formed on the surface of Ni foam, it is usually necessary to clean the Ni foam with acid to remove the inert layer before use. As a conventional treatment solution, hydrochloric acid (HCl) is widely used in the literature of Ni foam [6]. However, the HCl treatment can only be used to remove the oxide layer on the surface of the Ni foam without changing the surface roughness of the skeleton.

As early as 1959, W.J. Tegart in his book first proposed a mixed acid solution, glacial acetic acid ($CH_3COOH$) + nitric acid ($HNO_3$) + sulfuric acid ($H_2SO_4$) + phosphoric acid ($H_3PO_4$), to remove the inert layer on the surface of pure nickel electrode [14]. In 2012, Grdeń et al. first applied this mixed acid solution for the surface etching of Ni foam and found that this etching method can increase the electrochemically active and specific surface areas of Ni foams [15]. In 2019, Zhang et al. used mixed acid-etched Ni foam to the electrocatalytic oxidation of ethanol and obtained better performance [16].

Therefore, in order to explore the influence of the thickness of the Ni foam on cell performance and to fully exert the catalytic function of the catalyst even at a lower loading, in this chapter, firstly, the impact of Ni foam electrodes of different thicknesses on cell performance is discussed. Secondly, a mixed acid-etched nickel foam electrode is explored via the electrode characterization, electrochemical characterization to cell performance test, and the electrode with low palladium loading and high performance is achieved. This work shows the potential for an in-depth exploration of metal foam electrodes.

## 2. Effect of nickel foam thickness on the performance of an alkaline direct ethanol fuel cell

### 2.1 Experimental

In this experiment, Nafion 212 membrane is used as an electrolyte membrane, which conducts $K^+$ for internal ion conduction. The cell principle is shown in **Figure 1**.

#### *2.1.1 Materials and electrode preparation*

Three different thicknesses (0.3 mm, 0.6 mm, and 1.0 mm) of Ni foam (110 ppi, Tianyu Heze, China) were cut into 2.0 cm × 2.0 cm as electrode support materials. Before utilization, the Ni foam needs to be pretreated. First, immerse it in acetone for 20 min. Then, soak it in 1.0 M HCl for 10 min to remove the surface oxide, and finally, rinse it thoroughly with deionized water.

On the anode side, 20 wt.% Pd/C catalyst (Shanxi rock®, China) and 5 wt.% PTFE solution (Hesen®, China) were mixed in isopropanol for preparing the catalyst ink. After stirring the catalyst ink in the ultrasonic bath for 10 min, anode Ni foam electrodes of different thicknesses were prepared by the dip-coating method [11]. The Ni foam is firstly weighed with a scale, then directly immersed in the catalyst ink for 3 min, and finally dried with infrared light for 3 min, and then weighed again. The dipping-drying process was repeated till the 1.0 mg $cm^{-2}$ Pd loading was on the Ni foam to form an electrode.

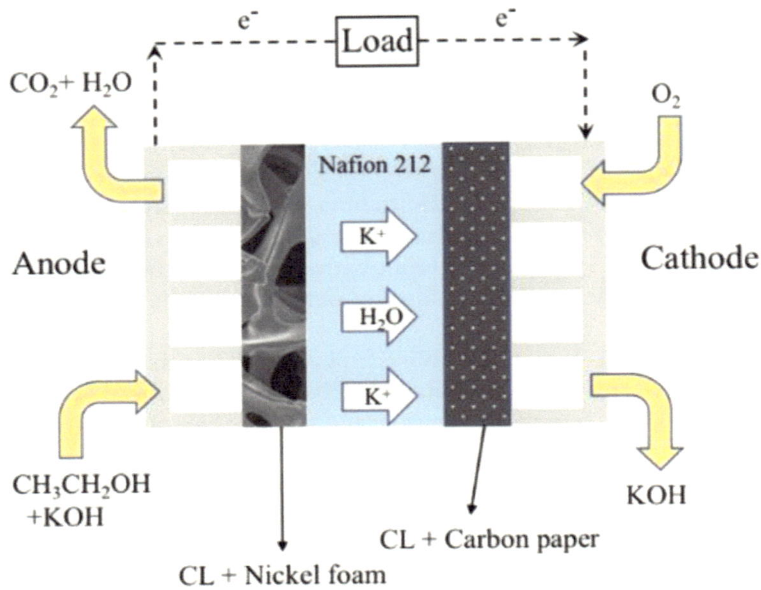


**Figure 1.**
*Schematic diagram of an alkaline direct ethanol fuel cell using Nafion 212 membrane.*

On the cathode, a commercial carbon paper electrode (Pt/C nanoparticles with a metallic loading of 2.0 mg $cm^{-2}$, purchased from Hesen®, China) was applied [17].

### 2.1.2 Electrode characterization

The X-ray diffraction (XRD) measurement was performed in the D8 ADVANCE diffractometer device (Bruker, Germany), and the Cu Kα source was operated at 40 kV and 40 mA. The data is collected in the 2θ range of 20–90° at 5° $min^{-1}$.

The surface morphology and electrode structure of Ni foam were characterized by a scanning electron microscope (SEM, Hitachi S-3400, Japan).

Cyclic voltammetry (CV) experiments were carried out using a three-electrode cell [18] in an electrochemical workstation (CHI 660E, China) to study the Electrochemical Specific Surface Area (ECSA) of the anode catalyst. Nickel foam electrode, graphite rod, and saturated calomel electrode (SCE) are used as working electrode, a counter electrode, and reference electrode, respectively. At room temperature, the CV curves were tested with a scan rate of 50 mV $s^{-1}$ in 1 M KOH solution saturated with $N_2$. The active area of the working electrode is 1.0 $cm^2$, and the loaded Pd amount is 1.4 mg.

CHI 660E is used to measure the electrochemical impedance spectroscopy of the ADEFC at 0.4 V [19]. The working temperature of the fuel cell is well controlled at 60°C.

### 2.1.3 Evaluation of cell performance

The cell performance was evaluated in the MEA composed of a self-made Ni foam anode, commercial carbon paper cathode, and Nafion 212 membrane. Nafion 212 membrane was immersed in 3 M KOH solution for 24 h to enhance the conductivity of $K^+$ ions and then washed with deionized water before using [20].

The Nafion 212 membrane was sandwiched between the anode and cathode, and a 2.0 cm × 2.0 cm MEA was assembled without hot pressing. The anode and cathode flow field plates with serpentine flow channels.

*DOI: http://dx.doi.org/10.5772/intechopen.100287*

During the performance test, the anode was fed with 3 M ethanol and 3 M KOH solution by a peristaltic pump with a flow rate of 1.0 ml min$^{-1}$, and the cathode was provided dry oxygen with the flow rate of 100 ml min$^{-1}$.

The cell temperature was controlled at 60°C by a heating rod. The Arbin BT-I fuel cell test system was used to test the performance of a single cell, and the discharge curve was obtained by controlling the current with regular increment.

## 2.2 Results and discussion

The structural information for clean Ni foam and the Pd/C-Ni foam electrodes were obtained by XRD, as shown in **Figure 2**. In **Figure 2a**, the diffraction pattern

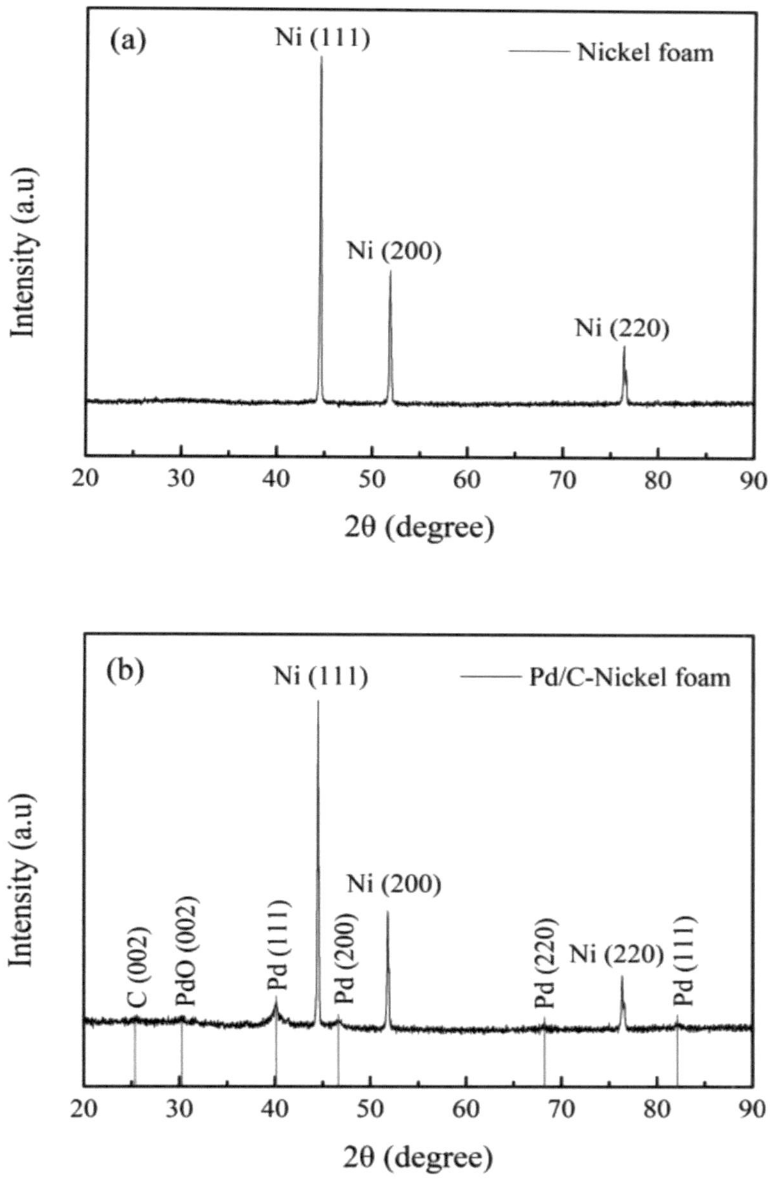


**Figure 2.**
*XRD pattern of (a) the clean Ni foam and (b) the Pd/C-Ni foam electrode.*

of a clean Ni foam has three peaks at the 2θ values of 44.50°, 51.85°, and 76.39°, relating with the (111), (200), and (220) planes of the Ni face-centered cubic structure (FCC). It shows that the surface of pretreated Ni foam is clean without impurities.

In **Figure 2b**, the Pd/C-Ni foam electrode exhibits four more peaks at the 2θ of 40.13°, 46.61°, 68.16°, and 82.09°, respectively, corresponding to the (111), (200), and (220) planes of Pd FCC structure. The XRD result indicates that the Pd catalyst is deposited on the Ni foam skeleton. It should be also mentioned that the carbon (002) peak at 25.56° is relatively weak, may be caused by the Pd/C catalyst is mainly distributed inside the three-dimensional structure of Ni foam, while XRD tests the surface of the foam, such that the peak obtained is not notable.

**Figure 3a** and **b** are the SEM images of clean Ni foam (0.6 mm) after pretreatment. It was found that the original Ni foam skeleton was even, and the porous network structure was regular.

From **Figure 3c**, the catalysts supported on the surface of Ni foam electrode with 0.3 mm thickness are massive and agglomerate severely, so that the catalyst particles wrapped inside do not participate in the catalytic reaction, the ECSA is reduced, and the chemical reaction kinetics weakens. The corresponding cross-sectional view, **Figure 3d** showed the accumulation of the catalyst more clearly and fails to show the advantages of the three-dimensional structure of Ni foam.

It can be seen from **Figure 3e** that the catalyst on the surface of 0.6 mm Ni foam electrode made full use of the supported noble metal catalyst and was evenly supported on the Ni foam skeleton. The even distribution of catalysts on Ni foam skeleton can also speed up the electron transfer. The electrons released by the ethanol oxidation reaction on the outer surface of the Pd catalyst can be fast transferred to the flow field plate through the Ni foam, reducing the electron transfer resistance. At the same time, it can be evidently seen from **Figure 3f** that catalyst loading density was moderate, the pores were more than the 0.3 mm Ni foam, and the mass transfer resistance was smaller.

Similar to that of the 0.6 mm electrode, **Figure 3g** and **h** show the images of the catalyst supported on the 1.0 mm Ni foam electrode. It is regarded that when the thickness of Ni foam exceeds 0.6 mm, the Pd catalyst with 1.0 mg $cm^{-2}$ loading will not have much agglomeration and the catalyst can be distributed evenly on the Ni foam skeleton. However, when the thickness of the Ni foam expands, the path length of the electron transfer will also be larger, which increases the transfer resistance [21].

**Figure 4** shows the cell performance of electrodes made of Ni foam with three different thicknesses. When different thicknesses of Ni foam were used, PTFE gaskets of corresponding thickness were used for sealing. As shown in **Figure 4** and **Table 1**, the maximum power densities for 0.3 mm, 0.6 mm, and 1.0 mm Ni foam electrodes were 23.2, 56.3, and 32.7 mW $cm^{-2}$, respectively. The 0.6 mm foam electrode held the highest power density, as 2.4 times higher than that of 0.3 mm and 1.8 times as that of 1.0 mm Ni foam. At the discharging voltage of 0.1 V, the corresponding maximum current densities were 178, 400, and 182 mA $cm^{-2}$, respectively.

In the low current density region (<50 mA $cm^{-2}$), as dominated by the activation polarization and was related to the active sites of catalyst, the cell voltage for 0.3 mm Ni foam electrode dropped sharply from open-circuit voltage (OCV) to 0.34 V. For the 0.6 mm and 1.0 mm thickness electrodes, the trend of voltage drop was quite similar and kept at 0.48 V and 0.46 V, respectively. From **Figure 3c** and **d**, the reason can be found, since 0.3 mm thickness is too thin to support the Pd loading of 1.0 mg $cm^{-2}$, thus causing catalyst agglomeration and large activation

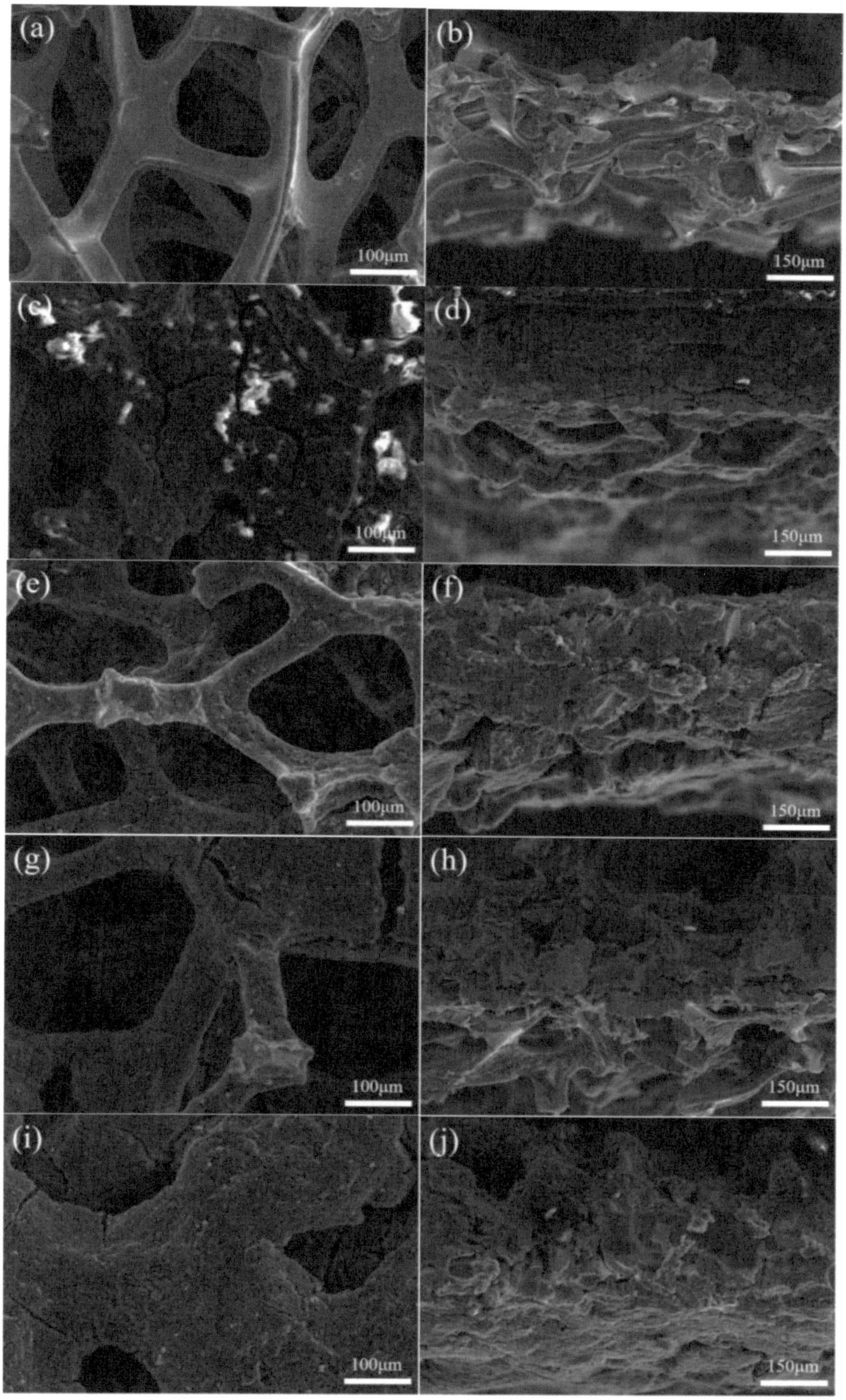


**Figure 3.**
*SEM images show the top view of Ni foam electrode with different thickness and different Pd loading. (a) 0.6 mm clean NF, (c) 0.3 mm NF with 1.0 mg cm$^{-2}$ Pd loading, (e) 0.6 mm NF with 1.0 mg cm$^{-2}$ Pd loading, (g) 1.0 mm NF with 1.0 mg cm$^{-2}$ Pd loading and (i) 0.6 mm NF with 2.0 mg cm$^{-2}$ Pd loading. While SEM images (b), (d), (f), (h) and (j) show the cross-sectional morphology of corresponding Ni foam electrode.*

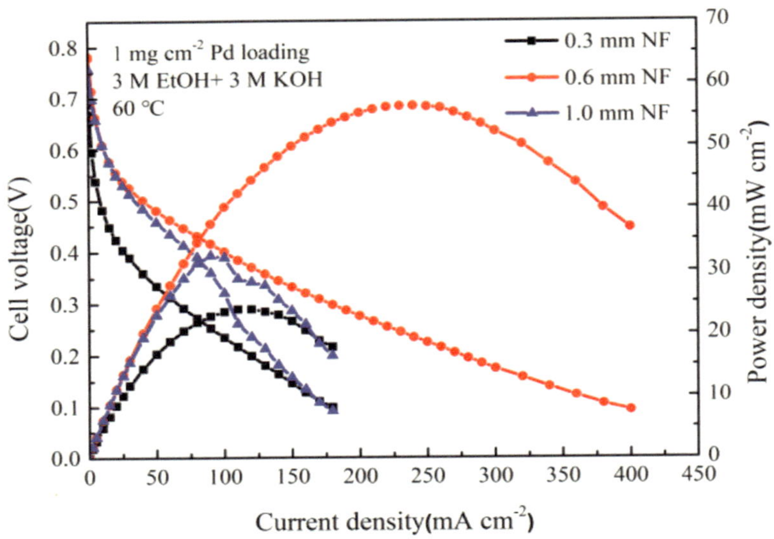


**Figure 4.**
*Cell performance comparison of ADEFC with different thickness of Ni foam electrodes.*

| Nickel foam thickness (mm) | OCV | 50 (mA cm$^{-2}$) | 100 (mA cm$^{-2}$) | 150 (mA cm$^{-2}$) |
|---|---|---|---|---|
| 0.3 | 0.67 V | 0.34 V | 0.24 V | 0.15 V |
| 0.6 | 0.78 V | 0.48 V | 0.40 V | 0.33 V |
| 1.0 | 0.76 V | 0.46 V | 0.32 V | 0.16 V |

**Table 1.**
*Corresponding to the voltage values of several current density points in the above polarization curve.*

polarization loss associated with low current densities. However, the electrode of 0.6 mm Ni foam still has enough permeability, which is conducive to the migration of reactant species.

Similarly, in the medium current density range (50–100 mA cm$^{-2}$) controlled by ohmic loss and was related to the resistance of each part of MEA, the voltage of 1.0 mm Ni foam dropped faster than that of the electrode with 0.6 mm Ni foam, which was from 0.48 V to 0.40 V and from 0.46 V to 0.32 V, respectively.

For the high current density region (>150 mA cm$^{-2}$), as controlled by concentration polarization and was mainly related to the mass transfer, only the voltage of 0.6 mm Ni foam electrode could still remain at 0.33 V, implying excellent mass transfer property. This is due to the larger the Ni foam thickness, the longer the electron conduction and mass transfer path, resulting in inferior performance.

In a word, there is an optimal choice between the thickness of the Ni foam electrode and the amount of catalyst loaded, and the thickness should be optimized by considering the trade-off between the electron conduction and mass transfer resistance.

In order to clarify the polarization performances of Ni foam electrodes with different thicknesses, EIS tests were taken as shown in **Figure 5**. The intersection between the high-frequency region and the X-axis represents the ohmic resistance (R1), which is the sum of the electrodes, membrane, and contact resistances. It is mainly dominated by the internal resistance of the membrane [22]. At 0.4 V, the R1 of the 0.3 mm, 0.6 mm and 1.0 mm Ni foam electrodes were 0.414 Ω cm$^2$,

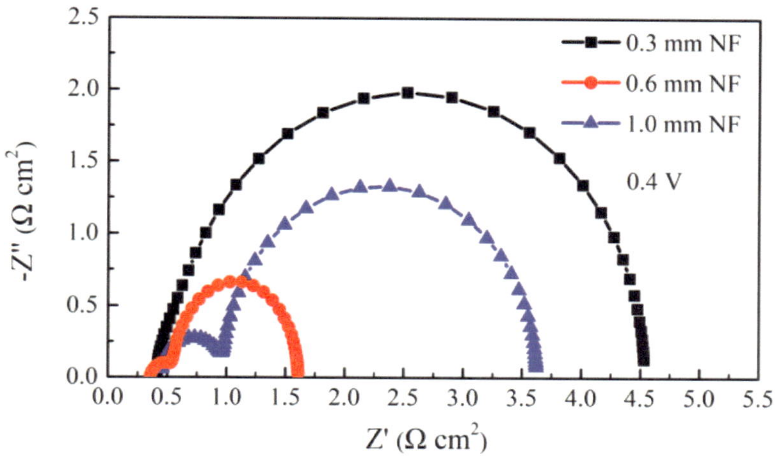


**Figure 5.**
*EIS comparison of ADEFC with different thickness of Ni foam electrodes at 0.4 V.*

0.352 Ω $cm^2$ and 0.445 Ω $cm^2$, respectively. The R1 of the 0.6 mm Ni foam electrode was slightly smaller, as the catalyst distribution on the Ni foam was the most uniform, and the contact between the membrane and the electrode was closer so that the contact resistance was smaller (see **Table 2**).

The value of the intersections of the high and low frequencies on the X-axis can be attributed to the total resistance of charge transfer (cathode and anode), which is also called the electrode reaction resistance (R2). As the exact same commercial cathode electrode is used for each test, the difference in R2 can reflect the difference in anode resistance [23]. The R2 values of the three different thickness Ni foam electrodes were 4.121 Ω $cm^2$, 1.227 Ω $cm^2$ and 3.177 Ω $cm^2$, respectively. It can be found that the better the cell performance, the smaller the electrode reaction resistance. This can be explained as follows: the three-dimensional space of the 0.3 mm Ni foam electrode is narrow, and the catalyst agglomeration is serious, which reduces the catalyst utilization, weakens the EOR kinetics, so that cause the increase of R2. For the 0.6 mm Ni foam electrode, the catalyst is uniformly loaded on the skeleton and almost fully utilized, which contributes to the rapid oxidation of ethanol and reduces the R2. Meanwhile, the 1.0 mm Ni foam electrode has a longer path for electron transfer, which increases the charge transfer resistance, so that a larger R2.

The cyclic voltammetry was used to study the catalytic activity of Ni foam electrodes with different thicknesses in the alkaline medium under the same Pd loading. For each CV curve in **Figure 6**, there was a hydrogen desorption peak at about 0.5 V in the forward potential scan, followed by an OH adsorption peak and an oxidation peak on Pd. In the backward direction scan, there was a reduction peak around 0.6 V, which relates to the reduction of PdO. The reduction current peak is

| Nickel foam thickness (mm) | R1 (Ω $cm^2$) | R2 (Ω $cm^2$) |
|---|---|---|
| 0.3 | 0.414 | 4.121 |
| 0.6 | 0.352 | 1.254 |
| 1.0 | 0.445 | 3.177 |

**Table 2.**
*EIS data of ADEFC with different thickness of Ni foam electrodes at 0.4 V.*

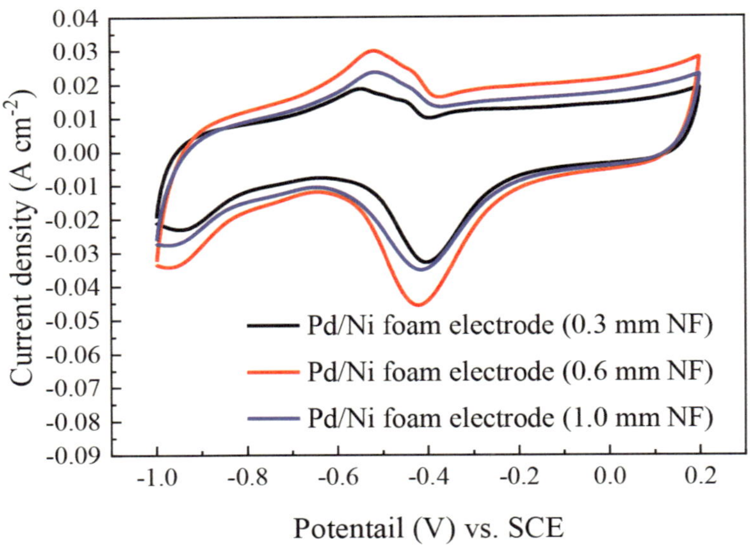


**Figure 6.**
*CV curves of the different thickness of Ni foam electrodes in 1 M KOH at a scan rate of 50 mV s$^{-1}$.*

proportional to the amount of OH adsorbed on the Pd surface, such that this can be applied to calculate the ECSA. The larger the ECSA, the more active sites the catalyst holds [24]. The calculation formula of ECSA (Eq. (1)) is as follows:

$$\mathrm{ECSA} = \frac{Q}{qm} \tag{1}$$

where, $Q$ is the coulombic charge obtained by integrating the PdO reduction peak area (μC), $m$ is the Pd loading on the electrode (mg), and $q$ is the charge value of the reduction of PdO of 405 μC cm$^2$. The ECSA values of 0.3 mm, 0.6 mm, and 1.0 mm Ni foam electrodes were 11 m$^2$ g$^{-1}$, 17 m$^2$ g$^{-1}$ and 15 m$^2$ g$^{-1}$, respectively, showing that the 0.6 mm Ni foam electrode has the highest ECSA, which can utilize the catalyst better to boost a higher cell performance.

# 3. A mixed acid-etched nickel foam electrode with ultra-low Pd loading for superior performance of an alkaline direct ethanol fuel cell

## 3.1 Experimental

### *3.1.1 Ni foam pretreatment*

Firstly, the Ni foam (0.6 mm, 110 ppi, purchased from Heze Tianyu Technology Development Co., Ltd., China) is immersed in acetone and sonicated for 20 min. Next, soak the Ni foam in 5 M HCl for 10 min. Finally, the Ni foam was cleaned with plenty of deionized water.

For comparison, another piece of Ni foam was soaked in anhydrous mixed acids solution (50 g of 99.5 wt.% $CH_3COOH$ + 45 g of 66.5 wt.% $HNO_3$ + 12 g of 85 wt.% $H_3PO_4$ + 10 g of 96.5 wt.% $H_2SO_4$) for 15 s, and the rest of the steps were the same.

### *3.1.2 Electrochemical characterization*

To explore the effect of different acid treatments to remove the Ni foam inert layer, cyclic voltammetry (CV) test was performed in a three-compartment electrochemical cell by an electrochemical workstation (CHI 604E, Shanghai Chenhua Instrument Co., Ltd., China). Cut a piece of 1 cm × 2 cm pretreated Ni foam, use pretreated Ni foam as the working electrode, platinum mesh as the counter electrode, and SCE as the reference electrode. All the potentials applied to refer to an RHE scale. The three electrodes were immersed in 1 M KOH solution, and the electrolyte solution was degassed by bubbling $N_2$ for 30 min.

To characterize the ECSA and EOR activity of the Pd/Ni foam electrode treated with different acids, a CV test was performed in a three-compartment electrochemical cell by the electrochemical workstation CHI 604E. Cut a piece of 1 cm × 2 cm pretreated Ni foam, and 100 μL 15 mM $Na_2PdCl_4$ solution was added dropwise to its surface, then 100 μL 5 mM $NaBH_4$ and 2.5 mM NaOH mixed solution was added dropwise. Thus, the Pd catalyst was chemically reduced on the Ni foam surface to form a Ni foam electrode (Pd/Ni foam) [18]. The area of catalyst deposition was only 1 cm × 1 cm, and the other 1 cm × 1 cm was the part of the working electrode used to collect current. When testing ECSA of the catalyst, the test solution was 1 M KOH. When testing the catalyst activity of EOR, the test solution was a mixed solution of 1 M KOH and 1 M ethanol.

### *3.1.3 Physical characterization*

To observe the surface microstructure, a scanning electron microscope (SEM, Hitachi S-3400, Japan) was tested on the pretreated Ni foam and the Pd/Ni foam electrode.

In order to detect the diffraction peak intensity of metallic Ni on the surface of foam, X-ray diffraction (XRD, Bruker, Germany) was tested on the pretreated Ni foam. Data were collected in the 2θ range 20–90 at 5° $min^{-1}$.

### *3.1.4 Cell performance test*

Two pieces of 2 cm × 2 cm Ni foam treated with different acids were cut and soaked in 1 mL 15 mM $Na_2PdCl_4$ solution for reaction. To make sure that the Ni foam surface was sufficiently replaced by Pd, the replacement time was set as 12 h. After drying and weighing, the Ni foam anode was prepared, and Pd loading was 0.35 mg $cm^{-2}$. The difference in electrode potential between metals was used to directly replace the catalyst from the precursor solution onto the surface of the Ni foam. The method was simple and no binder was needed [25].

A prepared Ni foam anode, a KOH pre-treated Nafion 212 membrane, and a commercial carbon-paper cathode (Pt/C nanoparticles with a metallic loading of 2.0 mg $cm^{-2}$, purchased from Hesen, China) were stacked together to form an MEA. The Nafion 212 membrane was immersed in 3 M KOH solution for 12 h to enhance the $K^+$ conductivity of the membrane, where $K^+$ realized internal loop circulation [26]. The MEA was placed in a cell fixture with a 2 cm × 2 cm serpentine flow channel made of stainless steel 316L, assembled into a single cell. The Arbin instrument (Arbin BT-I, USA) was used for cell performance testing. The anode was fed with a mixed solution of 3 M KOH and 3 M ethanol, and the peristaltic pump controlled the flow rate to 2 mL $min^{-1}$. The cathode was fed with normal concentration oxygen, and the rotor flowmeter controlled the flow rate to 100 mL $min^{-1}$. An external heating rod controlled the cell temperature to 60°C.

### 3.2 Results and discussion

It can be clearly seen from **Figure 7a** that the pristine Ni foam had a three-dimensional network structure with a disorderly arranged metal skeleton and high porosity. By observing **Figure 7b** and **c**, it was intuitively found that the HCl treatment method did not significantly change the Ni foam skeleton surface, and the metallic nickel layer was still smooth. However, there were obvious changes in the mixed acids treated Ni foam skeleton surface of **Figure 7d** and **e**, with many small holes of about 1 μm in diameter on the skeleton surface, which enhanced the roughness of Ni foam. Meanwhile, these micro-holes existed in the three-dimensional structure of the whole skeleton, which expanded the specific surface area of Ni foam.

The reason for the significant surface difference is that the reaction of Ni foam soaking in HCl solution is mild. HCl without strong oxidizing properties does not react strongly with the metallic nickel, so the surface of the skeleton remains smooth. Whereas, in anhydrous mixed acids solution, $CH_3COOH$ and $H_3PO_4$ are weak acids with the properties of acid, and $HNO_3$ and $H_2SO_4$ are strong acids with strong oxidizing and corrosive properties. Therefore, a short-term immersion treatment could cause a violent redox reaction with metallic nickel. Part of the metallic nickel on the skeleton surface was consumed by the reaction, and the gases produced by the reaction (e.g., $NO_2$ and $SO_2$) left dense holes in the skeleton. It should be noted that it was precise because of the strong corrosion ability of the mixed acids that the immersion time should be controlled in a very short time. In this experiment, the soaking time was controlled at 15 s, because the Ni foam was completely dissolved when the soaking time exceeded 1 min.

It can be seen from **Figure 8** that both the Ni foam treated with HCl and Ni foam treated mixed acids had three obvious diffraction peaks at 2θ values of 44.6°, 51.9° and 76.4°, corresponding to the standard diffraction peaks of metallic nickel on (111), (200) and (220) crystal planes (JCPDS. No 04-0850), which indicated that different acid treatment methods did not affect the metallic nickel lattice parameters.

However, by checking the intensity of the diffraction peaks of these two Ni foams, it was not difficult to find that the peak intensities of the (111) and (200) crystal planes of the Ni foam treated with HCl were higher than the corresponding peak intensities of the Ni foam treated with the mixed acids (the black dotted line

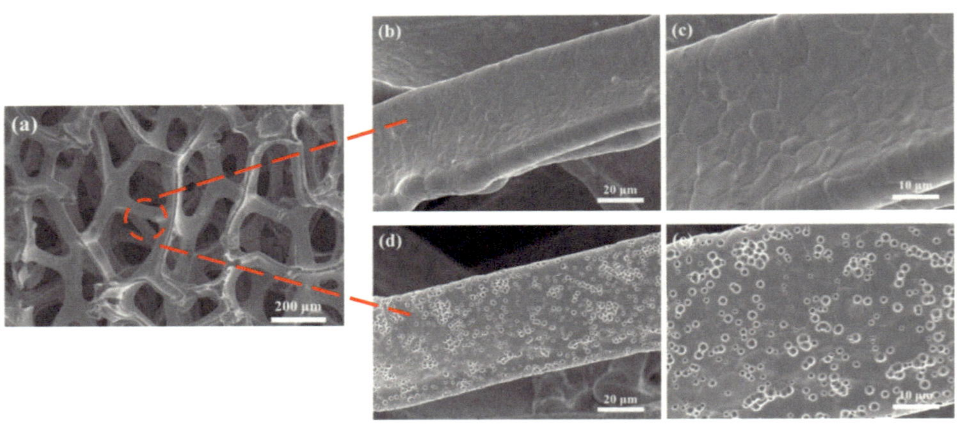


**Figure 7.**
*SEM images of (a) pristine Ni foam; (b) Ni foam skeleton treated with HCl; (c) an enlarged view of (b); (d) Ni foam skeleton treated with mixed acids; (e) an enlarged view of (d).*

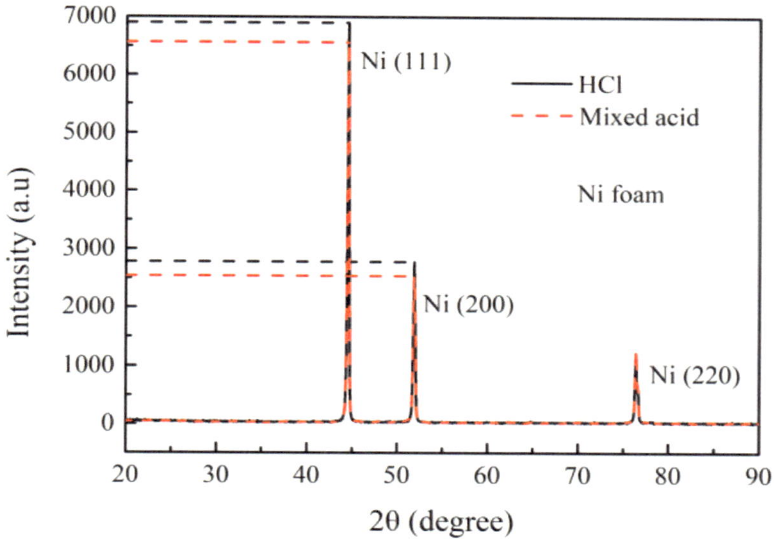


**Figure 8.**
*XRD patterns of Ni foam treated with HCl and Ni foam treated with mixed acids.*

above the red dotted line). The reason for this phenomenon is that the mixed acids consume part of metallic nickel during the etching process, resulting in a slight decrease in the X-ray diffraction intensity. It corresponds to the fact that the surface of Ni foam treated with HCl is smooth and neat, while the surface of Ni foam treated with mixed acid is rough, as shown in **Figure 7**.

**Figure 9** shows the standard CV curve of the Pd catalyst. In the potential range of 0.1–0.5 V, it corresponded to the hydrogen region. The oxidation peak was H desorption, OH adsorption, while the reduction peak was H adsorption; In the potential range of 1–1.2 V, it corresponded to the oxygen region, where PdO was

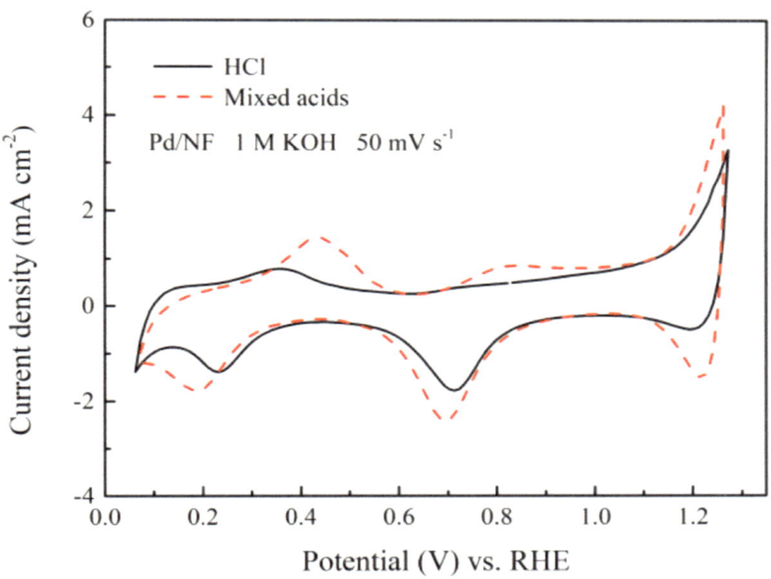


**Figure 9.**
*CV curves of Pd/Ni foam electrode treated with HCl and Pd/Ni foam electrode treated with mixed acids in 1 M KOH solution. Scan rate: 50 mV s$^{-1}$. Room temperature.*

gradually formed, and the corresponding peak at 0.7 V was the reduction peak of PdO. There was an obvious oxidation peak in the potential range of 0.3–0.5 V, and a reduction peak in the potential around 0.1 V, which correspond to the mutual oxidation-reduction between metal Ni and $\alpha$-$Ni(OH)_2$, respectively [27]. Also their peaks overlap with the H adsorption/desorption peaks. Generally, the PdO reduction peak was used to compare the electrochemically active area of the Pd catalyst [24].

It also can be seen from **Figure 9** that the Pd/Ni foam electrode treated with mixed acids had a larger ECSA. This result was attributed to a large number of micro-holes formed on the Ni foam skeleton after mixed acids etching. These holes existed greatly increase the specific surface area of Ni foam. When the Pd catalyst was loaded, some of the catalyst particles would be deposited in these holes, which improved the utilization rate of the catalyst. Therefore, the Pd catalyst loaded on the Ni foam treated with mixed acids exposed more active sites and had larger ECSA.

**Figure 10** shows the EOR electro-catalytic performance of Pd/Ni foam electrodes treated with different acids. It was found that the maximum current density of the Pd/Ni foam electrode etched by HCl under 0.77 V was only 32 mA $cm^{-2}$, while the EOR performance of the Pd/Ni foam electrode etched by mixed acids was much better, as the maximum current density measured at 0.85 V was 56 mA $cm^{-2}$. The catalytic performance of EOR has been greatly improved. This was due to the porosity in the nickel skeleton after being etched by the mixed acids, which increased the overall specific surface area. The Pd catalyst was fully utilized, the active sites exposed by the catalyst were more, and the ECSA was higher and the better EOR performance was obtained.

**Figure 11** exhibits the preparation process of the Pd/Ni foam electrode. From this figure, it can be seen that the electrode preparation method was simple, and the electrode can be made by immersing only 3 times. **Figures 12** and **13** are the SEM test and cell performance test of the Pd/Ni foam electrode prepared through the preparation process as shown in **Figure 11**.

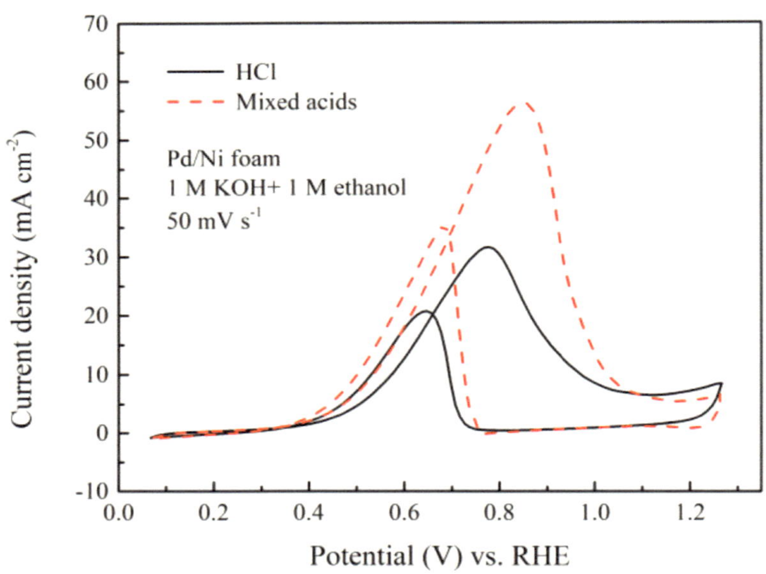


**Figure 10.**
*CV curves of Pd/Ni foam electrode treated with HCl and Pd/Ni foam electrode treated with mixed acids in 1 M KOH + 1 M ethanol. Scan rate: 50 mV $s^{-1}$. Room temperature.*

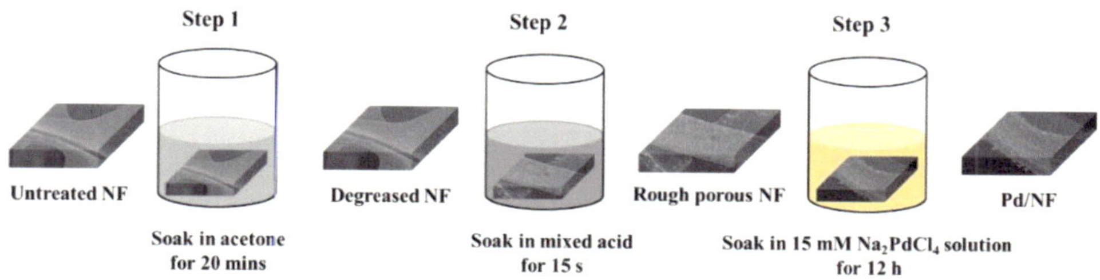


**Figure 11.**
*Preparation process of Pd/Ni foam electrode.*

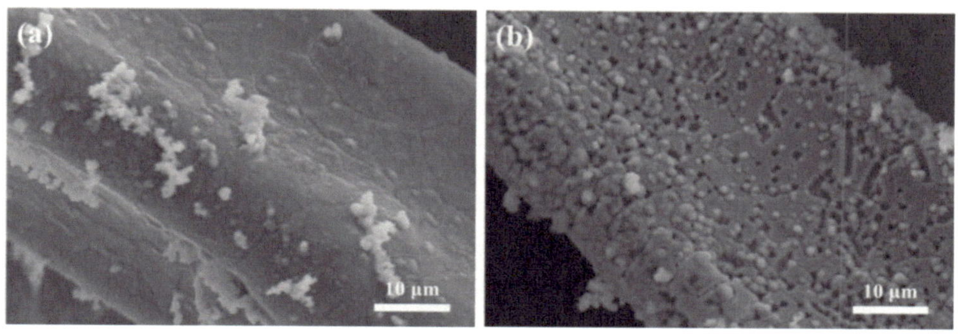


**Figure 12.**
*SEM images of (a) Pd/Ni foam electrode treated with HCl; (b) Pd/Ni foam electrode treated with mixed acids. The two electrodes were prepared by spontaneous deposition method as shown in* ***Figure 12****. Pd loading: 0.35 mg cm*$^{-2}$.

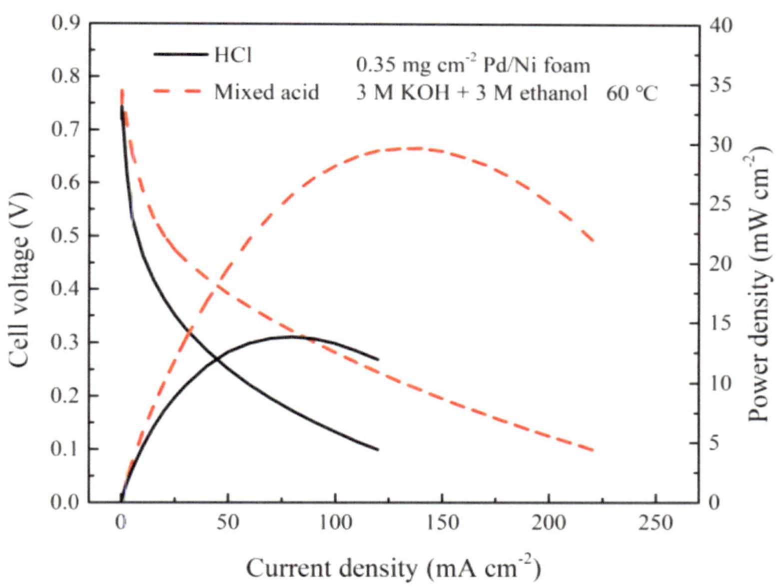


**Figure 13.**
*Polarization and power density curves of ADEFCs with different acid treated Pd/Ni foam anodes. Pd loading: 0.35 mg cm*$^{-2}$*. Test temperature: 60°C. Feeding solution: 3 M KOH and 3 M ethanol aqueous solution.*

It can be seen from **Figure 12a** that the Ni foam skeleton was smooth, and the Pd catalyst had uneven load and slight agglomeration on the skeleton surface, which showed that the conventional HCl treatment method reduced the catalyst

| Anode | OCV (V) | Maximum current density at 0.1 V (mA $cm^{-2}$) | Peak power density (mW $cm^{-2}$) |
|---|---|---|---|
| Pd/Ni foam treated with HCl | 0.719 | 120 | 15 |
| Pd/Ni foam treated with mixed acids | 0.747 | 220 | 30 |

**Table 3.**
*Important parameter values of cell performance corresponding to* ***Figure 13****.*

utilization. Compared with **Figure 12b**, it was found that the skeleton surface was rough and porous. Most of the Pd particles were uniformly filled in the holes left by the mixed acid etching. These small holes increased the active area of the Pd catalyst.

According to the detailed cell performance information in **Figure 13** and **Table 3**, it can be found that the OCV of the Ni foam electrode treated with HCl was 0.719 V, the corresponding maximum current density at 0.1 V was 120 mA $cm^{-2}$, and the peak power density was 15 mW $cm^{-2}$. While the mixed acid-etched Ni foam had excellent cell performance, OCV was 0.747 V, the maximum current density at 0.1 V was 220 mA $cm^{-2}$, and the peak power density was 30 mW $cm^{-2}$. Further analysis showed that the cell voltage of the HCl treated electrode dropped quickly from the low current density area. The voltage loss in this area was mainly the activation loss, which was determined by the kinetics of the catalytic oxidation of ethanol and the catalytic reduction of oxygen. Because of the smooth surface of the Ni foam electrode treated by HCl, the catalytic activity was low at low catalyst loading, leading to slow catalytic oxidation of ethanol and a sharp drop in voltage. The Ni foam skeleton etched by mixed acid was rough and porous, which greatly improved the utilization rate of the catalyst and had a high catalyst active area so that even with a small amount of catalyst, it could still exhibit high ethanol oxidation activity and excellent cell performance. This conclusion was also supported by the SEM images in **Figure 12**.

It is worth noting that the prepared electrode applied such a low catalyst loading (0.35 mg $cm^{-2}$ Pd) for single cell test, as to make full use of the advantage that the small holes created by the mixed acids treatment can effectively increase the specific surface area. If the catalyst loading was too high, the catalyst particles would be layered on the Ni foam skeleton. When the skeleton was layered, the catalyst particles filled in the holes would be covered by the upper catalyst layer, which not only failed to reflect the advantages of porosity, but also reduced the utilization of the catalyst inside the holes [28]. It was precise because even with low catalyst loading that higher cell performance was obtained, the superiority of the mixed acid treatment method could be better reflected. This result indicates the potential for in-depth exploration of metal foam electrodes.

## 4. Conclusions

This work firstly focuses on the effect of different thickness Ni foam anodes on the performance of alkaline direct ethanol fuel cells. The result exhibited that among the 0.3 mm, 0.6 mm, and 1.0 mm thickness Ni foam electrodes, the 0.6 mm thick electrode had the best cell performance, reaching a maximum power density of 56.3 mW $cm^{-2}$ at 60°C, 2.4 times higher than that of 0.3 mm and 1.8 times of that of 1.0 mm. The reason was that the 0.3 mm Ni foam was thin, the catalyst was prone to aggregate on the electrode surface, and was unevenly distributed in the three-dimensional space, leading to the increase of electrochemical reaction resistance and the decrease of performance. On the other hand, the electron and mass transfer channel of Ni foam with a thickness of 1.0 mm was long, which caused larger polarization. These results have been proven by SEM image, polarization curve test, EIS test, and CV test. It is necessary to optimize the thickness of the Ni foam to more effectively use the catalyst, balance the resistance of electron conduction and mass transfer, and improve the cell performance.

Secondly, the mixed acids etching method is applied to pretreat the Ni foam. It is found that etching can obtain porous skeleton surface, since the strong oxidizing and corrosive nature of the mixed acids can quickly consume the metallic nickel, leaving many micro-holes on the skeleton surface. Using the three-electrode system, it is concluded that the Pd/Ni foam electrode pretreated with mixed acids has a larger ECSA and higher ethanol oxidation activity than HCl. A mixed acid treated Ni foam anode with low Pd loading (0.35 mg $cm^{-2}$) is prepared by simply soaking three times for ADEFC performance testing. The peak power density reaches 30 mW $cm^{-2}$, which is double the performance of the HCl treated anode. The performance improvement is attributed to the micro-holes produced by mixed acids etching, which enhance the roughness of the skeleton and improve the ECSA of the catalyst. This work opens a new platform for in-depth exploration on metal foam electrodes.

## Acknowledgements

The work described in this chapter was fully supported by Grants from the NSFC, China (No. 51676092), State Key Laboratory of Engine at Tianjin University (No. K2020-14), Six-Talent-Peaks Project in Jiangsu Province (No. 2016-XNY-015), and High-Tech Research Key Laboratory of Zhenjiang City (No. SS2018002).

## Conflict of interest

The authors declare no conflict of interest.

## Author details

Qian Xu[1]*, Jiajia Zhang[1] and Chunzhen Yang[2]*

1 Institute for Energy Research, Jiangsu University, Zhenjiang, China

2 School of Materials Science and Engineering, Sun Yat-Sen University, Guangzhou, China

*Address all correspondence to: xuqian@ujs.edu.cn
and yangchzh6@mail.sysu.edu.cn

## References

[1] Badwal SPS, Giddey S, Kulkarni A, Goel J, Basu S. Direct ethanol fuel cells for transport and stationary applications—A comprehensive review. Applied Energy. 2015;**145**:80-103. DOI: 10.1016/j.apenergy.2015.02.002

[2] Ong BC, Kamarudin SK, Basri S. Direct liquid fuel cells: A review. International Journal of Hydrogen Energy. 2017;**42**:10142-10157. DOI: 10.1016/j.ijhydene.2017.01.117

[3] Fadzillah DM, Kamarudin SK, Zainoodin MA, Masdar MS. Critical challenges in the system development of direct alcohol fuel cells as portable power supplies: An overview. International Journal of Hydrogen Energy. 2019;**44**:3031-3054. DOI: 10.1016/j.ijhydene.2018.11.089

[4] An L, Zhao TS. Transport phenomena in alkaline direct ethanol fuel cells for sustainable energy production. Journal of Power Sources. 2017;**341**:199-211. DOI: 10.1016/j.jpowsour.2016.11.117

[5] Ozden A, Shahgaldi S, Li X, Hamdullahpur F. A review of gas diffusion layers for proton exchange membrane fuel cells—With a focus on characteristics, characterization techniques, materials and designs. Progress in Energy and Combustion Science. 2019;**74**:50-102. DOI: 10.1016/j.pecs.2019.05.002

[6] Tan WC, Saw LH, Thiam HS, Xuan J, Cai Z, Yew MC. Overview of porous media/metal foam application in fuel cells and solar power systems. Renewable and Sustainable Energy Reviews. 2018;**96**: 181-197. DOI: 10.1016/j.rser.2018.07.032

[7] Bazylak A, Sinton D, Liu ZS, Djilali N. Effect of compression on liquid water transport and microstructure of PEMFC gas diffusion layers. Journal of Power Sources. 2007;**163**:784-792. DOI: 10.1016/j.jpowsour.2006.09.045

[8] Kuruneru STW, Vafai K, Sauret E, Gu Y. Application of porous metal foam heat exchangers and the implications of particulate fouling for energy-intensive industries. Chemical Engineering Science. 2020;**228**:115968. DOI: 10.1016/j.ces.2020.115968

[9] Paserin V, Marcuson S, Shu J, Wilkinson DS. CVD technique for inco nickel foam production. Advanced Engineering Materials. 2004;**6**:454-459. DOI: 10.1002/adem.200405142

[10] Wang YL, Zhao YQ, Xu CL, Zhao DD, Xu MW, Su ZX, et al. Improved performance of Pd electrocatalyst supported on three-dimensional nickel foam for direct ethanol fuel cells. Journal of Power Sources. 2010;**195**:6496-6499. DOI: 10.1016/j.jpowsour.2010.04.025

[11] Li YS, Zhao TS. A high-performance integrated electrode for anion-exchange membrane direct ethanol fuel cells. International Journal of Hydrogen Energy. 2011;**36**:7707-7713. DOI: 10.1016/j.ijhydene.2011.03.090

[12] Sun X, Li Y, Li MJ. Highly dispersed palladium nanoparticles on carbon-decorated porous nickel electrode: An effective strategy to boost direct ethanol fuel cell up to 202 mW cm$^{-2}$. ACS Sustainable Chemistry & Engineering. 2019;**7**:11186-11193. DOI: 10.1021/acssuschemeng.9b00355

[13] Grdeń M, Jerkiewicz G. Influence of surface treatment on the kinetics of the hydrogen evolution reaction on bulk and porous nickel materials. Electrocatalysis. 2019;**10**:173-183. DOI: 10.1007/s12678-019-0506-6

[14] Tegart WJM. The electrolytic and chemical polishing of metals in research and industry. 2nd ed. New York: Pergamon Press; 1959. p. 139. DOI: 10.1002/maco.19600110932

[15] Grdeń M, Alsabet M, Jerkiewicz G. Surface science and electrochemical analysis of nickel foams. ACS Applied Materials & Interfaces. 2012;**4**:3012-3021. DOI: 10.1021/am300380m

[16] Zhang C, Li CF, Li GR. Pd nanoparticles supported on the etched Ni foams as high-performance electrocatalysts for direct ethanol fuel cells. Journal of Electrochemistry. 2019;**25**:571-578. DOI: 10.13208/j.electrochem.181144

[17] Sun W, Zhang W, Su H, Leung P, Xing L, Xu L, et al. Improving cell performance and alleviating performance degradation by constructing a novel structure of membrane electrode assembly (MEA) of DMFCs. International Journal of Hydrogen Energy. 2019;**44**:32231-32239. DOI: 10.1016/j.ijhydene.2019.10.113

[18] Li Y, He Y. Layer reduction method for fabricating Pd-coated Ni foams as high-performance ethanol electrode for anion-exchange membrane fuel cells. RSC Advances. 2014;**4**:16879-16884. DOI: 10.1039/c4ra01399a

[19] Wu J, Yuan XZ, Wang H, Blanco M, Martin JJ, Zhang J. Diagnostic tools in PEM fuel cell research: Part I electrochemical techniques. International Journal of Hydrogen Energy. 2008;**33**:1735-1746. DOI: 10.1016/j.ijhydene.2008.01.013

[20] Hou H, Wang S, Jin W, Jiang Q, Sun L, Jiang L, et al. KOH modified Nafion112 membrane for high performance alkaline direct ethanol fuel cell. International Journal of Hydrogen Energy. 2011;**36**:5104-5109. DOI: 10.1016/j.ijhydene.2010.12.093

[21] Tseng CJ, Tsai BT, Liu ZS, Cheng TC, Chang WC, Lo SK. A PEM fuel cell with metal foam as flow distributor. Energy Conversion and Management. 2012;**62**:14-21. DOI: 10.1016/j.enconman.2012.03.018

[22] Yang SH, Chen CY, Wang WJ. An impedance study of an operating direct methanol fuel cell. Journal of Power Sources. 2010;**195**:2319-2330. DOI: 10.1016/j.jpowsour.2009.10.066

[23] Guo J, Chen R, Zhu F-C, Sun S-G, Villullas HM. New understandings of ethanol oxidation reaction mechanism on Pd/C and $Pd_2Ru/C$ catalysts in alkaline direct ethanol fuel cells. Applied Catalysis B: Environmental. 2018;**224**:602-611. DOI: 10.1016/j.apcatb.2017.10.037

[24] Ma L, Chu D, Chen R. Comparison of ethanol electro-oxidation on Pt/C and Pd/C catalysts in alkaline media. International Journal of Hydrogen Energy. 2012;**37**:11185-11194. DOI: 10.1016/j.ijhydene.2012.04.132

[25] Niu X, Zhao H, Lan M. Palladium deposits spontaneously grown on nickel foam for electro-catalyzing methanol oxidation: Effect of precursors. Journal of Power Sources. 2016;**306**:361-368. DOI: 10.1016/j.jpowsour.2015.12.044

[26] Zakaria Z, Kamarudin SK, Timmiati SN. Membranes for direct ethanol fuel cells: An overview. Applied Energy. 2016;**163**:334-342. DOI: 10.1016/j.apenergy.2015.10.124

[27] van Drunen J, Kinkead B, Wang MCP, Sourty E, Gates BD, Jerkiewicz G. Comprehensive structural, surface-chemical and electrochemical characterization of nickel-based metallic foams. ACS Applied Materials & Interfaces. 2013;**5**:6712-6722. DOI: 10.1021/am401606n

[28] Zhang J, Zhang W, Xing L, Su H, Xu Q. Experimental investigation on the effect of mixed acids etched nickel foam electrode on performance of an alkaline direct ethanol fuel cell. In: 2020 5th International Conference on Advances in Energy and Environment Research (ICAEER 2020); 18-20 September 2020; Virtual, Online. Vol. 194. China: EDP Sciences; 2020. p. 02021

Chapter 3

# Investigation of Synthesis Methods for Improved Platinum-Ruthenium Nanoparticles Supported on Multi-Walled Carbon Nanotube Electrocatalysts for Direct Methanol Fuel Cells

*Adebare Nurudeen Adewunmi, Sabejeje Akindeji Jerome, Su Huaneng and Lindiwe Eudora Khotseng*

## Abstract

This book chapter reports on various catalyst synthesis methods (impregnation, polyol, modified polyol, and microwave-assisted modified polyol methods) to determine which method would result in the most electrochemically active platinum-ruthenium (PtRu) electrocatalyst supported on multi-walled carbon nanotubes (MWCNTs) for methanol oxidation reaction in an acidic medium. Different techniques were used to characterize the synthesized catalysts, including the high-resolution transmission electron microscope used for morphology and calculating particle sizes, and X-ray diffraction for determining crystalline sizes. The electroactive catalyst surface area, ECSA of the electrocatalysts was determined using cyclic voltammetry (CV), while the electroactivity, electron kinetics, and stability of the electrocatalysts towards methanol oxidation were evaluated using CV, electrochemical impedance spectroscopy, and chronoamperometry, respectively. The microwave-assisted modified polyol method produced the PtRu/MWCNT electrocatalyst with the most enhanced electrocatalytic activity compared to other PtRu/MWCNT catalysts produced by the impregnation, polyol, and modified polyol methods.

**Keywords:** methanol oxidation, catalysts, multi-walled carbon nanotubes, electrocatalytic activity

## 1. Introduction

Catalyst synthesis methods have an influence on mean particle size, particle size distribution, the bulk and surface of catalysts' composition, the oxidation state of catalysts, the extent of catalyst alloying, the distribution of catalyst crystal surfaces, and catalyst morphology [1, 2], and hence on the catalytic activity of the metal catalysts [3]. The standard by which high-performance catalysts are evaluated includes a uniform composition in the entire nanoparticles, a complete alloying degree, a narrow nanoscale size distribution, and high dispersion on carbon support [4]. Various methods for the synthesis of fuel cell catalysts have been reported in the literature, including micro-emulsion, sputtering, and co-precipitation methods. In this study, we report on the synthesis of catalysts using the impregnation, polyol, modified polyol, and microwave-assisted modified polyol methods.

### 1.1 Impregnation method

The impregnation approach is frequently utilized for the manufacture of Pt-based catalysts. It is a straightforward chemical preparation process for catalyst synthesis that can create tiny particles in the 3–7 nm range with regulated loading [5, 6]. The impregnation method for the synthesis of platinum-ruthenium (PtRu) includes an impregnation step in which Pt and Ru precursor salts are mixed with the support material, which is typically high-surface-area porous or nanostructured carbon and penetrates into pores. The catalyst support aids in the penetration and wetting of the precursor and the carbon support confines the particle size growth during the reduction step. The chemical reduction can be carried out in the liquid phase with a reductive agent such as $Na_2S_2O_3$, $NaBH_4$, $Na_4S_2O_5$, $N_2H_4$ or formic acid, or in the gas phase with a reductive agent such as a flowing hydrogen stream at elevated temperatures. The difficulty in adjusting nanoparticle size and distribution is a key limitation of the impregnation process [5, 6]. It has been noted that impregnated catalysts have a tendency to generate inhomogeneous agglomerations of active species at the support boundary, resulting in large-sized particles [1].

Other difficulties include the use of chloride precursors, which could result in chloride poisoning and decreased catalytic activity and stability of the chloride-salt-produced catalyst. Metal nitrate/nitrite salts such as $Pt(NH_3)_2(NO_2)_2$ and RuNO $(NO_3)x$ [7], carbonyl complexes such as $Ru_3(CO)_{12}$ [8], and metal sulfite salts such as $Na_6Pt(SO_3)_4$ and $Na_6Ru(SO_3)_4$ [9] as metal precursors for Pt and Ru, respectively, have been investigated for impregnation methods that could use chloride-free precursors. When compared to the traditional Cl-containing route, these chloride-free pathways provide improved dispersion and catalytic activity. In this chapter, we report on an impregnation method where $NaBH_4$ was used as the reducing agent and ethylene glycol (EG) as the solvent.

### 1.2 Polyol reduction method

The polyol method includes the following common steps: (1) preparation of Pt-containing colloids; (2) deposition of the colloids onto the support, and (3) chemical reduction of the mixture. The synthesis occurs in an organic or aqueous medium where the metal precursor is reduced chemically in the presence of a protective agent (i.e., NR41, $PPh_3$, PVP, SB12, or PVA). Other colloid methods using several reducing agents, organic stabilizers, or shell-removing approaches have also been developed in recent years. The catalyst is supported with catalyst support to enhance the surface area and the dispersion of the catalyst. To achieve a limited size distribution, the colloidal metal nanoparticles are stabilized by steric hindrance or electrostatic charges. Coating the metal core with organic chain molecules can offer steric stability [10, 11]. The aggregation of charged colloids or adsorbed ions is limited by the electrostatic repulsion of similar charges. The use of protective agents, which may influence the catalytic activity of the nanoparticles, poses a problem for the polyol process, but it may be removed by washing in a suitable solvent or breakdown at temperatures in an inert atmosphere. There are also other challenges facing the polyol method, such as that it is time-consuming, complex, and expensive, which causes difficulty in terms of scaling up. The colloidal method prepares catalysts with nanoparticle size and narrow size distribution.

The polyol method that was employed in this study has been extensively explored as a preparation method for Pt [5, 12].

### 1.3 Modified polyol reduction method

Fievet et al. [13] pioneered the use of EG as both a solvent and a reducing agent. They found that EG may support colloidal metal particles in solution, resulting in a well-distributed solution. EG has a relatively high viscosity, and therefore it prevents Pt from being delivered to reaction sites too quickly, resulting in reduced Pt particle sizes [14]. This method is called the modified polyol method [15]. The modified polyol method is able to effectively synthesize very small and well-dispersed metal nanoparticles [2]. However, the synthesis parameters such as the water: EG ratio, the concentration of EG, and the pH of the solution have a great effect on the characteristics of the results [2]. Bimetallic catalysts, metal oxides, and metal sulfides with narrow particle size distributions, controlled compositions, and alloy structures have also been effectively prepared using the modified polyol technique [16]. The modified polyol approach, which uses EG as a reducing agent and solvent, was also used to make catalysts in this study. EG was utilized as both a reducing agent and a solvent for the Pt and Ru precursors in this method. The solution of EG, Pt, and Ru precursor salts was heated to 120–170°C during the reduction phase. EG is decomposed in this step, resulting in the reducing species ($CH_3CHO$-acetaldehyde, Eq. (1)) [17].

$$CH_3OCH_2OH \rightarrow CH_3CHO + H_2O \tag{1}$$

$$2CH_3CHO + (PtCl_6) + 6OH^- \rightarrow 2CH_3COO^- + Pt + 6Cl^- + 4H_2O \tag{2}$$

As represented in Eq. (2), acetaldehyde converts Pt ions into metallic Pt particles. The main feature of this polyol synthesis is that the acetate can act as a stabilizer for Pt and Ru colloids by forming chelate-type complexes via its carbonyl group. It is therefore unnecessary to use stabilization agents to prevent PtRu particles from agglomerating. As a result of using modified polyol synthesis, carbon-supported catalysts with reduced noble metal sizes and narrow size distribution are achieved.

### 1.4 Microwave-assisted modified polyol methods

The microwave synthetic approach was one of the methods employed in this study. A modified polyol method has been reported for this technique, with the deposition and reduction steps taking place in a microwave reactor. Microwave synthetic methodology has been utilized to manufacture catalysts because of its fast, uniform, homogeneous, and instant heating environment, which resulted in rapid reduction and facilitated metal particle nucleation [8, 18]. Microwave heating is a promising technology, with its applications rapidly growing due to its advantages over conventional heating, such as rapid volumetric heating, which increased reaction rates and shortened reaction time; however, to induce crystallization, a post-synthesis heat treatment was required [3]. Under such conditions, a microwave-assisted synthesis method is an appropriate option, with the added benefits of narrow size distribution and high purity [4]. At high pH conditions for depositions, promising results were reported with an average Pt size of 2.7 nm. However, because the reaction takes place in a closed system, the pH cannot be controlled throughout the duration of the reaction, hence, the entire scope of the reaction under these conditions is unknown [19].

## 2. Results and discussion

### 2.1 Physical characterization of the prepared electrocatalysts

X-ray diffraction (XRD) and the high-resolution transmission electron microscope (HRTEM) were used to learn more about the catalyst structure. The XRD patterns were performed with the Bruker AXS D8 advance equipment with Cu-K radiation and a wavelength of 1.5406 nm. With a scanning step of 0.035°, the Bragg angle range was 2θ = 10–90°. The instrumental contribution to peak profile characteristics was determined using a standard α-$Al_2O_3$ sample. A JEOL 2010 TEM system operating at 200 kV was used to obtain the HRTEM micrographs. The HRTEM samples were made by dispersing the carbon-supported electrocatalysts in ethanol and then casting a drop of the suspension onto a Cu-grid covered in carbon film for analysis. For each electrocatalyst, the particle size determined by HRTEM was obtained using Image J software over multiple areas.

### 2.2 XRD analysis

XRD measurements were performed to obtain the crystallographic information of the prepared catalysts. The XRD results presented in **Figure 1** indicate that all prepared metal catalysts presented a typical face-centered cubic crystallographic structure of PtRu crystals. The Bragg angles indicate that varying bimetallic interactions or alloying occurred in the PtRu crystals due to different catalyst preparation methods. The diffraction peak for carbon is at about 2θ = 25°. Other peaks are at 2θ values of 39.9°, 46.21°, 67.8°, and 81.2°, which are indexed to (111), (200), (220), and (311) planes of PtRu/MWCNT crystal structure, respectively. The strongest and sharpest diffraction peak for all four samples is at around 2θ = 39.9° indexed as (111) reflection of PtRu/MWCNT crystal planes prepared through the modified polyol and polyol methods, while the other characteristic PtRu/MWCNT diffraction peaks at 2θ of about 46.21°, 67.8°, and 81.2° corresponded to (200), (220), and

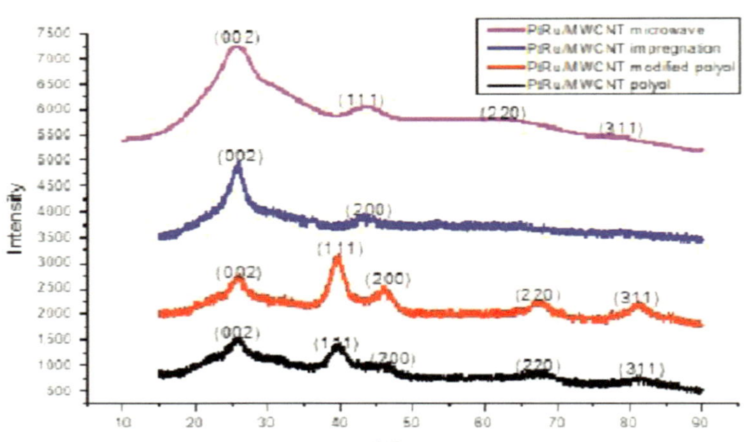


**Figure 1.**
*XRD spectra of PtRu electrocatalysts supported on multi-walled carbon nanotubes (MWCNTs) prepared through the impregnation, polyol, modified polyol, and microwave-assisted modified polyol methods.*

(311), respectively. Similar results were also reported by Kim et al. [20] and Wu et al. [21].

The crystalline size of the metal particles is calculated using Debye-Scherrer's equation, Kα/βCosθ, where K, Scherrer constant = 0.9, α, X-ray wavelength = 0.154 nm, and β(2θ), the width of the diffraction peak (rad). From **Table 1** the average particle size can be seen, with PtRu/MWCNT nanoparticles prepared through microwave-assisted modified polyol having the smallest crystalline size of 1.95 nm, followed by PtRu/MWCNT nanoparticles prepared through the modified polyol method with a crystalline size of 4.33 nm.

Catalyst nanoparticles are the dark dots as shown in the HRTEM micrographs in **Figure 2**. PtRu/MWCNT nanoparticles are well distributed with little agglomeration. PtRu/MWCNT electrocatalysts prepared through the microwave-assisted modified polyol method has the least agglomeration, followed by PtRu/MWCNT modified polyol when compared with other electrocatalysts evident from their higher electroactive catalyst surface areas of $4.15 \times 10^2$ $m^2/g$ and $3.2 \times 10^2$ $m^2/g$, respectively an advantage for enhanced electrocatalytic activities. Both of these catalysts gave the smallest particle sizes of 1.87 and 4.14 nm, respectively. The other electrocatalysts PtRu on MWCNT support particle sizes were between 5.77 and 6.90 nm. The particle sizes of the electrocatalysts obtained were comparable with the average particle sizes following the order PtRu/MWCNT impregnation > PtRu/MWCNT polyol > PtRu/MWCNT modified polyol > PtRu/MWCNT microwave-assisted modified polyol.

In **Figure 2** the histograms reveal the mean particle sizes and their nanoparticle size distributions for all PtRu/MWCNT electrocatalysts prepared through all four synthesis methods; however, the two electrocatalysts prepared through the impregnation method and polyol method exhibited better distribution of nanoparticles as compared to other electrocatalysts. The particle size was determined using Image J software estimated from 50 particles selected randomly from HRTEM micrographs of the PtRu/MWCNT nanoparticles. The mean particle sizes of PtRu/MWCNT impregnation, PtRu/MWCNT polyol, PtRu/MWCNT modified polyol, and PtRu/MWCNT microwave-assisted modified polyol were 6.90, 6.51, 4.14, and 1.87 nm, respectively.

## 2.3 Electrochemical characterization of the electrocatalysts

Electrochemical measurements were carried out at ambient temperatures using a three-electrode configuration, which includes a working electrode, a counter electrode, and a reference electrode. An Ag/AgCl electrode as a reference electrode and a Pt foil of a large area as a counter electrode were used. The working electrode was a glassy carbon disc (5 mm in diameter with a geometric area of 0.196 $cm^2$) covered with a thin layer of catalyst of fine film. Before the experiment, the electrode substrate was pre-treated by polishing it with a 0.05 μm $Al_2O_3$ particle suspension on a moistened microcloth. All the electrochemical experiments were carried out, namely electrochemical impedance spectroscopy (EIS),

| PtRu electrocatalysts | Crystalline size (nm) | Particle size (nm) |
|---|---|---|
| PtRu/MWCNT modified polyol | 4.33 | 4.14 |
| PtRu/MWCNT polyol | 6.75 | 6.51 |
| PtRu/MWCNT impregnation | 7.11 | 6.90 |
| PtRu/MWCNT microwave | 1.95 | 1.87 |

**Table 1.**
*Properties of the PtRu electrocatalysts.*

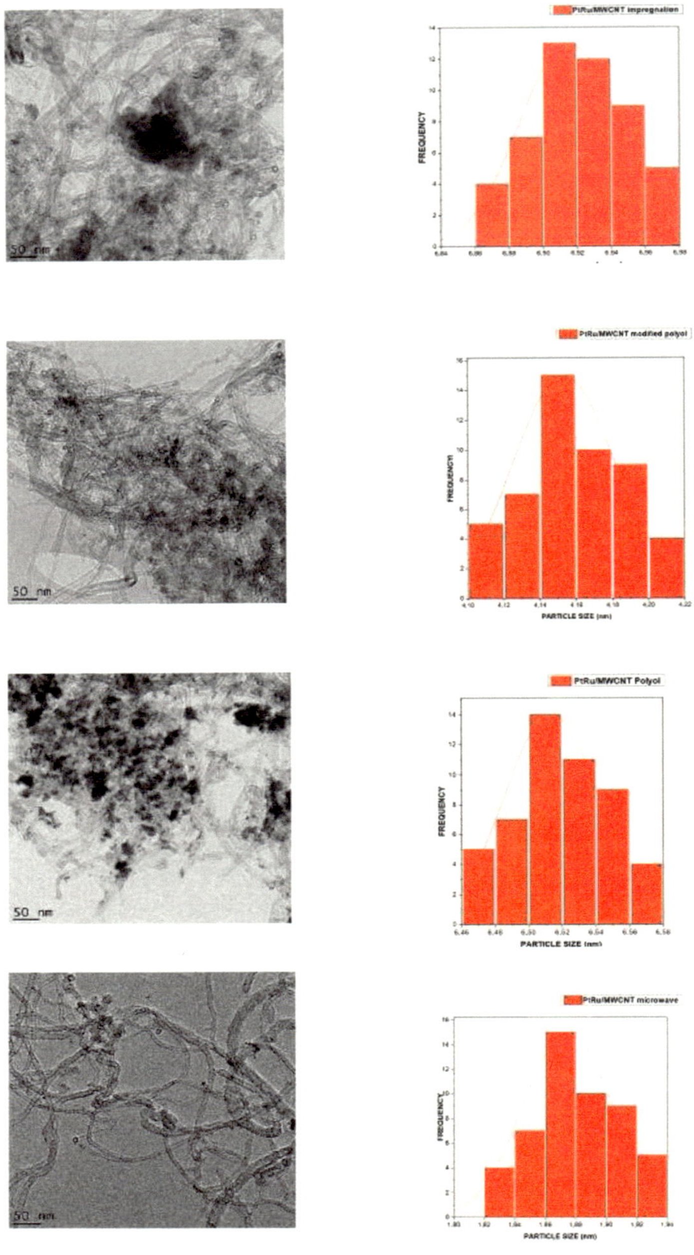


**Figure 2.**
*HRTEM images with their respective histograms for PtRu/MWCNT prepared through the impregnation, polyol, modified polyol, and microwave-assisted modified polyol methods.*

chronoamperometry (CA), and cyclic voltammetry (CV), and performed on an autolab electrochemical workstation (PGSTAT128N, Eco Chemie, the Netherlands). CA tests were carried out for the electrocatalytic stability of the PtRu/MWCNT catalysts for the methanol electro-oxidation. The CA was carried out for 30 minutes. CV evaluations were carried out at 30 mV/s covering a potential window from −0.2 to 1.2 V vs. Ag/AgCl. Perchloric acid was used as the electrolyte. Inert nitrogen gas was used to deaerate the solutions. To obtain a homogeneous catalyst layer, a stock solution was first prepared by mixing 20 ml of isopropanol, 79.6 ml of ultra-pure water, and 0.4 ml of 5wt% Nafion solution in a 100-ml volumetric flask. Thereafter, 10 mg of the catalyst powder was measured into a 10-ml vial and 5 ml of stock solution was added, mixed thoroughly, and sonicated for 60 minutes in an ultrasonicator. A measured volume of this mixture was dropped on top of the glassy carbon disc and then dried to form the desired catalyst layer.

Electrochemical activities of the prepared catalysts in a 0.5 M $HClO_4$ solution were firstly examined by CV. From the CV of the prepared electrocatalysts, the adsorption peaks for the different catalysts were observed. The peak area of the adsorption peak of the electrocatalysts in the CV was used to determine the electroactive surface area of the catalysts using the equation 2 [22]:

$$ECSA = \frac{Q}{210\mu C/cm2.m.Ag} \quad (3)$$

where $Q$ is the charge from the adsorption peak in Coulomb taking within the negative potential region of $-0.2$ to 0.08 V in the forward scan, as shown in **Figure 3**, $m$ is the working electrode Pt loading in mg $cm^{-2}$, $Ag$ is the geometric surface area of the electrode (5 mm in diameter, Ag = 0.196 $cm^2$) and 210 $\mu Ccm^{-2}$ is the value for the charge of full coverage for a clean polycrystalline Pt monolayer [23].

The obtained ECSA values were $4.15 \times 10^3$ $cm^2/g$ for PtRu/MWCNT microwave-assisted modified polyol, $3.2 \times 10^3$ $cm^2/g$ for PtRu/MWCNT modified polyol, $0.35 \times 10^3$ $cm^2/g$ for PtRu/MWCNT polyol and $0.28 \times 10^2$ $m^2/g$ for PtRu/MWCNT impregnation. The higher ECSA value of PtRu/MWCNT prepared through the microwave-assisted modified polyol method can be attributed to its lower particle size value, hence a higher surface area compared to the other electrocatalysts.

### 2.4 Methanol oxidation

The electrocatalytic activity of the PtRu/MWCNT series catalysts towards the methanol oxidation was investigated using the CV technique in 0.5 M $HClO_4$ with 2 M methanol at a scan rate of 30 $mVs^{-1}$, as shown in **Figure 4**.

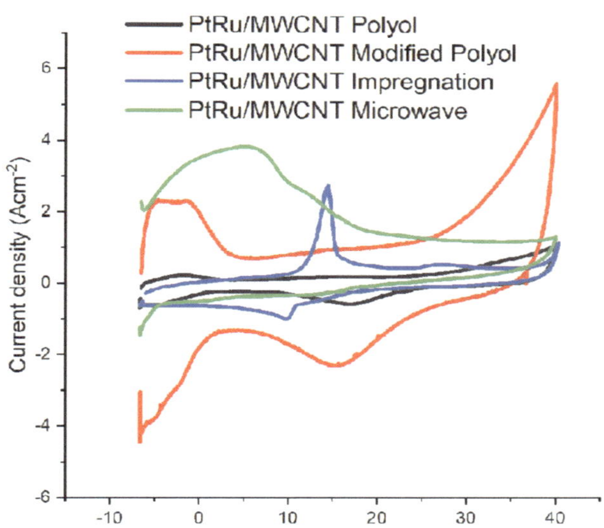


**Figure 3.**
*Cyclic voltammograms of PtRu/MWCNT electrocatalysts in $N_2$-saturated 0.5 M perchloric acid $HClO_4$ at a scan rate of 30 $mV.s^{-1}$.*

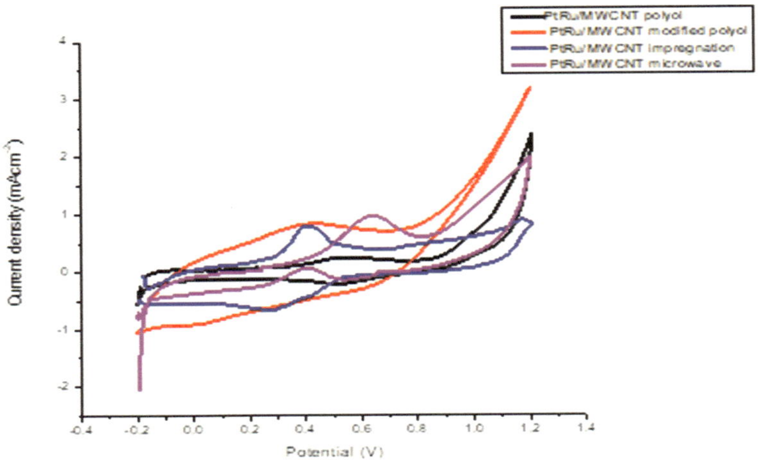


**Figure 4.**
*Cyclic voltammograms of PtRu/MWCNT electrocatalysts in $N_2$-saturated 0.5 M perchloric acid $HClO_4$ and 2 M methanol at a scan rate of 30 $mV.s^{-1}$.*

| PtRu electrocatalysts | $E_{(onset)}$ [a] [V vs. Ag/AgCl] | $E_f$ [b] [V vs. Ag/AgCl] | Ij [c] ($mAcm^{-2}$) | Electroactive catalyst surface area ($m^2/g$) |
|---|---|---|---|---|
| PtRu/MWCNT modified polyol | −0.194 | 0.415 | 0.16 | $3.2 \times 10^2$ |
| PtRu/MWCNT polyol | −0.187 | 0.522 | 0.041 | $0.35 \times 10^2$ |
| PtRu/MWCNT impregnation | −0.151 | 0.411 | 0.188 | $0.28 \times 10^2$ |
| PtRu/MWCNT microwave | −0.198 | 0.633 | 0.190 | $4.15 \times 10^2$ |

*[a] Onset potential, [b] forward anodic peak potential at 30 $mVs^{-1}$, [c] forward anodic peak current density at 30 $mVs^{-1}$.*

**Table 2.**
*Comparison of the electrocatalytic activity of the catalysts for methanol oxidation.*

The electrocatalytic activity towards methanol oxidation is summarized in **Table 2**. By comparing the characteristics of the CVs, the change in catalyst preparation methods leading to varying compositions of Pt and Ru in the metal alloys was found to substantially enhance the catalytic activity for methanol electro-oxidation. First, the onset potentials (a measure of catalytic activity) of methanol oxidation for the PtRu/MWCNT prepared through the microwave-assisted modified polyol method and PtRu/MWCNT showed relatively lower values than that of PtRu/MWCNT electrocatalysts prepared through the impregnation, polyol and modified polyol methods. The positions of the onset potentials follow the order of PtRu/MWCNT microwave-assisted modified polyol < PtRu/MWCNT modified polyol < PtRu/MWCNT polyol < PtRu/MWCNT impregnation. Second, the forward peak current densities (measure of the maximum catalyst performance) of the PtRu/MWCNT catalysts took the order PtRu/MWCNT microwave-assisted modified polyol > PtRu/MWCNT impregnation > PtRu/MWCNT modified polyol > PtRu/MWCNT polyol. Therefore, PtRu/MWCNT prepared through the microwave-assisted modified polyol method exhibited the most prominent electrochemical performance in terms of the highest forward peak current density and the lowest onset potential, followed by PtRu/MWCNT prepared through the modified polyol method.

It was found that PtRu/MWCNT produced through the microwave-assisted modified polyol method of Pt: Ru ratio close to 1:1 outperformed all other PtRu/MWCNT electrocatalysts produced through other synthesis methods in methanol electro-oxidation reaction evident from the current density of 0.190 mA/cm$^2$.

Electro-oxidation of methanol to form $CO_2$ can be via dual path mechanisms consisting of non-CO and adsorbed CO reactive intermediates [24]:

$$Pt(CH_3OH)ads + H_2O \rightarrow CO_2 + 6H + +6e \quad (4)$$

$$Pt(CH_3OH)ads \rightarrow Pt(CO)ads + 4H + +4e \quad (5)$$

The non-CO reaction pathway is preferred for methanol oxidation for which it does not involve CO, a poison for Pt metal. The adsorbed CO reaction pathway often presents, however, in which the intermediates via (CO)ads are mostly in the form of linearly bonded CO, that is, Pt = C = O [25]. Interaction of this complex on the catalyst surface leads to CO poisoning. The presence of Ru in the bimetallic catalyst assists in the oxidation of CO through chemisorbed -OH on the Ru sites [26]:

$$Ru + H_2O \rightarrow Ru - OH + H + +e \quad (6)$$

$$Ru\ OH + Pt(CO)ads \rightarrow Ru + Pt + CO2 + H + +e \quad (7)$$

In this way, the poisoned Pt is regenerated and can again participate in the oxidation of methanol. Due to the single species of CO and OH on Pt and Ru, respectively, the best results can be obtained when the Pt to Ru atomic ratio is 1:1 [27] (**Figure 5**).

The EIS technique was used to investigate the catalytic reaction kinetics for the methanol oxidation on the anodic PtRu/MWCNT electrocatalysts surfaces. The charge transfer resistance (Rct) values using equivalent circuit fitting were 5.985, 8.926, 4.061, and 6.184 kΩ for PtRu/MWCNT modified polyol, PtRu/MWCNT polyol, PtRu/MWCNT impregnation, and PtRu/MWCNT microwave-assisted modified polyol, respectively, indicating that PtRu/MWCNT prepared through the

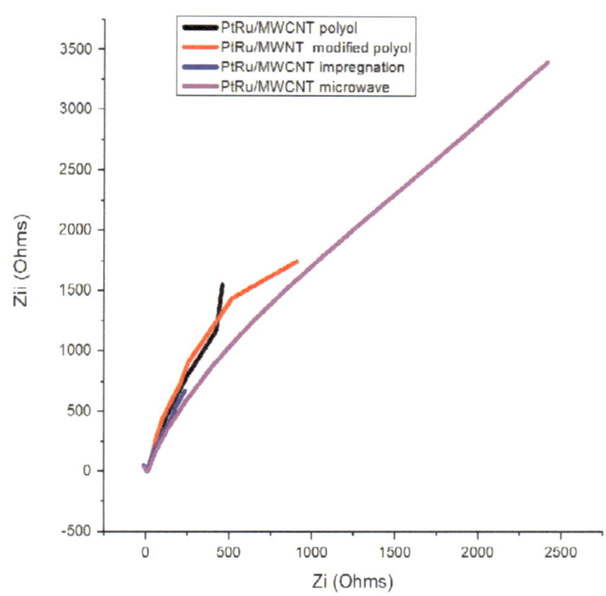


**Figure 5.**
*Electrochemical impedance curves of methanol oxidation on PtRu/MWCNT electrocatalysts prepared through different synthesis methods in $N_2$-saturated 0.5 M $HClO_4$ and 0.2 M methanol.*

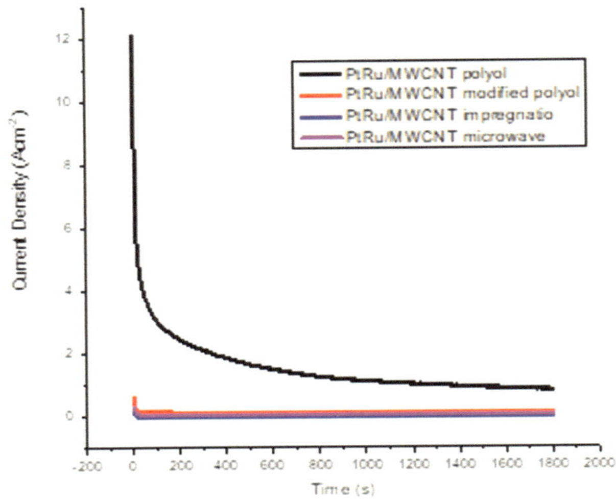


**Figure 6.**
*Chronoamperometry curves of methanol oxidation on PtRu/MWCNT electrocatalysts in 0.5 M $HClO_4$ and 2.0 M $CH_3OH$.*

impregnation method exhibited the best kinetics towards the methanol electro-oxidation with the least resistance to flow of electric current. PtRu/MWCNT prepared through the modified polyol method also showed promising kinetics with an Rct value of 5.985 kΩ.

The stability of the electrocatalysts is extremely important for their real applications in direct methanol fuel cells. **Figure 6** shows the CA of PtRu electrocatalysts on MWCNT support in $N_2$-saturated 0.5 M $HClO_4$ with 2.0 M methanol. This was to test the stability of the different catalysts after 1800 seconds. As observed at the start of the CA curve, the current density decreases sharply with time (I proportional to $t^{-1/2}$). The decreasing rate with time may characterize the inhibition of the electrodes by the methanol oxidation reaction products. When comparing the prepared catalysts, PtRu/MWCNT catalyst prepared through the polyol method performed better, followed by PtRu/MWCNT modified polyol. PtRu/MWCNT prepared by microwaving also showed better stability with higher current density than PtRu/MWCNT prepared through the impregnation method.

## 3. Conclusion

In this study, PtRu, supported by MWCNT, was successfully fabricated using the impregnation, polyol, modified polyol, and microwave-assisted modified polyol catalyst preparation methods. The synthesized electrocatalysts had crystalline sizes of 1.95–7.11 nm and average particle sizes of 1.87–6.90 nm, determined using XRD and HRTEM, respectively. The PtRu alloy phase is pronounced for the prepared electrocatalysts according to XRD analysis. It is found that the PtRu/MWCNT electrocatalyst produced through the microwave-assisted modified polyol method and PtRu/MWCNT modified polyol showed enhanced electrocatalytic activity towards methanol oxidation compared to other PtRu electrocatalysts on MWCNT support. Furthermore, the microwave-assisted prepared PtRu/MWCNT electrocatalyst had the largest current density for methanol oxidation compared to other electrocatalysts. This can be attributed to it having the smallest particle size and being the most active toward anode oxidation reaction. From the EIS, it was concluded that the PtRu/MWCNT electrocatalysts produced through the impregnation method exhibited a faster electrochemical reaction kinetics than both PtRu/MWCNT electrocatalysts produced through the polyol and modified polyol methods. Microwave-assisted modified polyol method PtRu electrocatalysts had the highest ECSA values compared to all other PtRu catalysts on MWCNT support, followed by PtRu/MWCNT produced by the modified polyol method. This was as a result of their smaller crystalline particle sizes of 1.95 and 4.33 nm, respectively. Polyol method synthesized PtRu/MWCNT was found to be the most stable electrocatalyst, followed by PtRu/MWCNT produced through the modified polyol method, as revealed by the chronoamperometry tests.

Based on all the results acquired in this investigation, it was concluded that the microwave-assisted modified polyol process of catalyst preparation method produced the best PtRu electrocatalyst on MWCNT to support the improved catalytic activity.

## Author details

Adebare Nurudeen Adewunmi[1]*, Sabejeje Akindeji Jerome[1], Su Huaneng[2] and Lindiwe Eudora Khotseng[1]*

1 Faculty of Natural Sciences, Department of Chemistry, University of the Western Cape, Bellville, South Africa

2 Institute for Energy Research, Jiangsu University, Zhenjiang, China

*Address all correspondence to: 3875793@myuwc.ac.za; lkhotseng@uwc.ac.za

## References

[1] Zhenlü W, Qingsheng L, Jianfeng Y, Tonghao W, Guojia W. Surface structure and catalytic behavior of silica-supported coppercatalysts prepared by impregnation and sol–gel methods. Applied Catalysis. 2003;**239**: 87-94. DOI: 10.1016/S0926-860X(02) 00421-0

[2] Lim B, Kim JW, Hwang SJ, Yoo SJ, Cho EA, Lim T-H, et al. Fabrication and characterization of high activity Pt/C electrocatalysts for oxygen reduction. Bulletin of the Korean Chemical Society. 2010;**31**(6):1577. DOI: 10.5012/ bkcs.2010.31.6.1577

[3] Krishnakumara T, Nicola P, Prasanna K, Perumala K, Jayaprakash R. Microwave-assisted synthesis and characterization of tin oxide nanoparticles. Materials Letters. 2008; **62**:3437-3440. DOI: 10.1016/j. matlet.2008.02.062

[4] Aarti SB, Denthaje KB, Cheuk-Wai T, Mysore SS. Microwave-assisted synthesis and magnetic studies of cobalt oxide nanoparticles. Materials Chemistry and Physics. 2011;**125**: 347-350. DOI: 10.1016/j. matchemphys.2010.11.003

[5] Dutrow BL, Clark CM. X-ray Powder Diffraction (XRD): Geochemical Instrumentation and Analysis. Available from: https://serc.carleton.edu/ research_education/geochemsheet/ techniques/XRD.html

[6] Kumar SMS, Hidyatai N, Herero JS, Irusta S, Scott K. Efficient tuning of the Pt nano-particles mono-dispersion on Vulcan XC-72R by selective pre-treatment and electrochemical evaluation of hydrogen oxidation and oxygen reduction reactions. International Journal of Hydrogen Energy. 2011;**36**: 5453-5465. DOI: 10.1016/j. ijhydene.2011.01.124

[7] Takasu Y, Fujiwara T, Murakami Y. Effect of structure of carbon-supported PTRU electrocatalysts on the electrochemical oxidation of methanol. Journal of the Electrochemical Society. 2000;**147**:4421-4427. DOI: 10.1149/ 1.1394080

[8] Dickinson AJ, Carrette LPL, Collins JA, Friedrich KA, Stimming U. Preparation of a Pt-Ru/C catalyst from carbonyl complexes for fuel cell applications. Electrochimica Acta. 2002;**47**:3733. DOI: 10.1016/S0013-4686(02)00343-2

[9] Friedrich KA, Geyzers KP, Dickinson AJ, Stimmin U. Fundamental aspects in electrocatalysis: From the reactivity of single-crystals to fuel cell electrocatalysts. Journal of Electroanalytical Chemistry. 2002;**261**: 524-525. DOI: 10.1016/S0022-0728(02) 00731-3

[10] Bonnemann H, Richards RM. Nanoscopic metal particles-synthetic methods and potential applications. European Journal of Inorganic Chemistry. 2001;**10**:2455

[11] Roucoux A, Schulz J, Patin H. Reduced transition metal colloids: A novel family of reusable catalysts? Chemical Reviews. 2002;**102**:3757. DOI: 10.1021/cr010350j

[12] Tang X, Li Y, Xu Y, Xin Q, Shen W. Structural features and catalytic properties of Pt/$CeO_2$ catalyst prepared by the modified reduction-deposition techniques. Catalysis Letters. 2004;**97**: 163-169

[13] Fievet F, Lagier JP, Figlarz M. Preparing monodisperse metal powders in micrometer and submicrometer sizes by the polyol process. MRS Bulletin. 1989;**14**:29

[14] Santiago EI, Varanda LC, Villulas HM. Carbon-supported Pt–Co

catalysts prepared by a modified polyol process as cathodes for PEM fuel cells. The Journal of Physical Chemistry. 2007;**111**:3146. DOI: 10.1021/jp0670081

[15] Available from: http://www.strem.com/uploads/

[16] He B, Chen Y, Liu H, Liu Y. Synthesis of solvent-stabilized colloidal nanoparticles of platinum, rhodium, and rutheniumby microwave-polyol process. Journal of Nanoscience and Nanotechnology. 2005;**5**:266

[17] Chiang YC, Lin WH, Chang YC. The influence of treatment duration on multi-walled carbon nanotubes functionalized by $H_2SO_4/HNO_3$ oxidation. Applied Surface Science. 2011;**257**:2401-2410

[18] Machida K, Fukuoka A, Ichikawa M, Enyio M. Preparation of platinum cluster-derived electrodes from metal carbonyl complexes and their electrocatalytic properties for anodic oxidation of methanol. Journal of the Electrochemical Society. 1958;**1991**:138

[19] Knupp S, Li W, Paschos O, Murray T, Snyder J, Haldar P. The effect of experimental parameters on the synthesis of carbon nanotube/nanofiber supported platinum by polyol processing techniques. Carbon. 2008;**46**:1276

[20] Kim Y, Kim HW, Lee S, Han J, Lee D, Kim JR, et al. The role of ruthenium on carbon-supported PtRu catalysts for electrocatalytic glycerol oxidation under acidic conditions. ChemCatChem. 2017;**9**(9):1683-1690. DOI: 10.1002/cctc.201601325

[21] Wu B, Zhu J, Li X, Wang X, Chu J, Xiong S. PtRu nanoparticles supported on p-phenylenediamine-functionalized multiwalled carbon nanotubes: Enhanced activity and stability for methanol oxidation. Ionics. 2019;**25**:181-189. DOI: 10.1007/s11581-018-2590-7

[22] Garsany Y, Baturina OA, Swider-Lyons KE, Kocha SS. Experimental methods for quantifying the activity of platinum electrocatalysts for the oxygen reduction reaction. Analytical Chemistry. 2010;**82**:6321-6328

[23] Vielstich W, Gasteiger HA, Lamm A, editors. Handbook of Fuel Cells—Fundamentals, Technology and Applications. Vol. 2. New York, NY: Wiley; 2003. p. 316. ISBN 0 471 49926 9; TRN: GB0400273

[24] Herrero E, Chrzanowski W, Wieckowski A. Dual path mechanism in methanol electrooxidation on a platinum electrode. The Journal of Physical Chemistry. 1995;**99**:10423. DOI: 10.1021/j100025a054

[25] Manohara R, Goodenough JB. Methanol oxidation in acid on ordered NiTi. Journal of Materials Chemistry. 1992;**2**:875. DOI: 10.1039/JM9920200875

[26] Meriaudeau P, Ellestad OH, Dufaux M, Naccache C. Metal-support interaction. Catalytic properties of $TiO_2$-supported platinum, iridium, and rhodium. Journal of Catalysis. 1982;**75**:243. DOI: 10.1016/0021-9517(82)90206-8

[27] Arico AS, Srinivasan S, Antonucci V. DMFCs: From fundamental aspects to technology development. Fuel Cells. 2001;**1**:133. DOI: 10.1002/1615-6854(200107)1:2<133::AID-FUCE133>3.0.CO;2-5

Section 2

# Electrocatalysts for a Cleaner Environment

Chapter 4

# Heterogeneous Electrocatalysts for $CO_2$ Reduction to Value Added Products

*M. Amin Farkhondehfal and Juqin Zeng*

## Abstract

The $CO_2$ that comes from the use of fossil fuels accounts for about 65% of the global greenhouse gas emission, and it plays a critical role in global climate changes. Among the different strategies that have been considered to address the storage and reutilization of $CO_2$, the transformation of $CO_2$ into chemicals and fuels with a high added-value has been considered a winning approach. This transformation is able to reduce the carbon emission and induce a "fuel switching" that exploits renewable energy sources. The aim of this chapter is to categorize different heterogeneous electrocatalysts which are being used for $CO_2$ reduction, based on the desired products of the above mentioned reactions: from formic acid and carbon monoxide to methanol and ethanol and other possible by products. Moreover, a brief description of the kinetic and mechanism of the $CO_2$ reduction reaction) and pathways toward different products have been discussed.

**Keywords:** $CO_2$ electroreduction, electrocatalyst, faradaic efficiency, metal-based

## 1. Introduction

Nowadays, global warming and $CO_2$ emissions as well as atmospheric $CO_2$ concentration are central topics in politics and scientific debate. The global energy supply based on fossil fuels has reached an unprecedented scale leading to excess anthropogenic $CO_2$ emission. $CO_2$ accumulates in the atmosphere and its concentration has surpassed 409 ppm in 2019 much higher than the 270 ppm during the pre-industrial era [1]. As a well-known greenhouse gas, accumulated $CO_2$ traps more infrared radiation, breaking the energy balance on the earth's surface. Using $CO_2$ as feedstock to produce valuable carbon-based chemicals is considered to be a feasible approach to close the carbon cycle and mitigate the climate change. Many strategies have been developed for $CO_2$ valorisation, including thermochemical, photochemical, electrochemical and biological approaches [2–5]. Among these methods, electrochemical conversion presents several advantages. Firstly, this method can use green chemicals as electrolytes and electricity from renewable energy sources, thus not contributing to new $CO_2$ emissions while transforming it [6]. Secondly, the products and conversion rates can be tuned by utilizing different catalysts and applying various potentials [7, 8]. Finally, the electrolyzer and electrolysis process for $CO_2$ conversion can be developed based on the already existing technologies such as water electrolyzers, polymer electrolyte membrane fuel cells, solid oxide fuel cells and so on [9]. However, the $CO_2$ reduction reaction ($CO_2$RR) involves several proton-assisted multiple-electron-transfer processes with similar standard potentials (V *vs* the reversible hydrogen electrode (RHE), Reactions (1)-(6) [10], leading to the formation of carbon monoxide (CO), formic acid (HCOOH), methane ($CH_4$), ethylene ($C_2H_4$), methanol ($CH_3OH$) and ethanol ($C_2H_5OH$). Moreover, hydrogen ($H_2$) evolution is the competing reaction in aqueous solution (Reaction (7)). Therefore, it is a challenge to control the selectivity of the $CO_2$RR from the thermodynamic view.

$$CO_2 + 2H^+ + 2e^- \rightarrow CO + H_2O\ E = -0.105\ V \tag{1}$$

$$CO_2 + 2H^+ + 2e^- \rightarrow HCOOH\ E = -0.169\ V \tag{2}$$

$$CO_2 + 8H^+ + 8e^- \rightarrow CH_4 + 2H_2O\ E = +0.169\ V \tag{3}$$

$$2CO_2 + 12H^+ + 12e^- \rightarrow C_2H_4 + 4H_2O\ E = +0.079\ V \tag{4}$$

$$CO_2 + 6H^+ + 6e^- \rightarrow CH_3OH + H_2O\ E = +0.017\ V \tag{5}$$

$$2CO_2 + 12H^+ + 12e^- \rightarrow C_2H_5OH + 3H_2O\ E = +0.08\ V \tag{6}$$

$$2H^+ + 2e^- \rightarrow H_2\ E = 0.000\ V \tag{7}$$

From the kinetic point of view, it is even more challenging to form chemical bonds for the complex and energetic molecule products [11]. Transferring one electron to the adsorbed $CO_2$ molecule to activate it (generating the radical $CO_2^{*-}$) is believed to be the rate-determining step of the $CO_2$RR on transition metal-based catalysts because of the high activation barrier needed for this step [12]. Consequently, much more negative potentials than the standard ones are needed to drive the $CO_2$RR. Therefore, an appropriately designed catalyst is essential in order to activate the $CO_2$ molecules. Once $CO_2^{*-}$ forms on the catalyst's surface, its reactivity in this state controls the distribution of final products. Both early and later studies [13–15] of electrochemical $CO_2$RR on various metal-based electrodes found that the radical $CO_2^{*-}$ interacts with the surface of the catalyst in different ways, depending on the intrinsic electronic surface's properties of the material. Hence, a suitable catalyst is necessary in order to selectively drive the $CO_2$RR and to obtain a specific product. In the present chapter, numerous electrocatalysts are classified based on the $CO_2$RR product, involving the reaction pathways and mechanism study.

## 2. CO-selective catalysts

CO is an important product from the reduction of $CO_2$ since it has high relevance for the chemical industry [16]. It is considered the most important $C_1$-building block and is intensively used in large industrial processes such as Fischer-Tropsch synthesis of hydrocarbons and Monsanto/Cativa acetic acid synthesis. By a techno-economic analysis that takes into consideration the costs of $CO_2$, electricity, separation, capital and maintenance, operation and the known product selectivity and outputs the levelized cost of the chemical produced, CO is one of the most economically viable and atom-economic targets [17].

In recent years, great efforts have been dedicated to the study of electrocatalysts for the electrochemical $CO_2RR$ to CO. **Table 1** summarizes the most widely investigated types.

| Electrocatalyst | Electrolyte | Potential (V vs. RHE) | Faradaic efficiency (%) | Current density ($-mA\ cm^{-2}$) | Reference |
|---|---|---|---|---|---|
| Au Nanoparticles | 0.5 M $KHCO_3$ | −0.67 | 90 | 5 | [18] |
| Au needles | 0.5 M $KHCO_3$ | −0.35 | 95 | 15 | [19] |
| Ag nanowires | 0.5 M $KHCO_3$ | −0.6 | 90 | 4.9 | [20] |
| Nanoporous Ag | 0.5 M $KHCO_3$ | −0.6 | 92 | 18 | [21] |
| Ag-$TiO_2$ | 0.1 M $KHCO_3$ | −1.3 | 35 | 30 | [20] |
| Zn nanoparticles | 0.1 M $KHCO_3$ | −0.95 | 98.1 | 17.5 | [22] |
| ZnO particles | 0.1 M $KHCO_3$ | −0.89 | 68 | 3.0 | [8] |
| Cu-Sn foam | 0.1 M $KHCO_3$ | −0.8 | 93 | 6.7 | [23] |
| Cu-Sn | 0.1 M $KHCO_3$ | −0.6 | 90 | 1.0 | [24] |
| Cu-$SnO_2$ | 0.5 M $KHCO_3$ | −0.7 | 93 | 4.6 | [25] |
| $Zn_{94}Cu_6$ foam | 0.5 M $KHCO_3$ | −0.95 | 90 | 8.0 | [26] |
| CuO-$Sb_2O_3$ | 0.1 M $KHCO_3$ | −0.8 | 90 | 5.0 | [27] |
| $SnO_x$/Ag | 0.5 M $KHCO_3$ | −0.6 | 85 | 0.7 | [28] |
| AuCu | 0.1 M $KHCO_3$ | −0.8 | 50 | — | [29] |
| Mn-N-C | 0.1 M $KHCO_3$ | −0.6 | ~80 | < 5 | [30] |
| Fe-N-C | | | | | |
| MnFe-N-C | | | | | |
| Ni-N-C | 0.1 M $KHCO_3$ | −1.0 | 95 | 15 | [31] |
| Ni-N-C | 0.1 M $KHCO_3$ | −0.81 | 80 | 13 | [32] |
| $FeN_4$/C | 0.1 M $KHCO_3$ | −0.6 | 93 | 1.5 | [33] |
| $FeN_5$/Graphene | 0.1 M $KHCO_3$ | −0.46 | 97 | 1.8 | [34] |
| $Fe^{3+}$–N–C | 0.5 M $KHCO_3$ | −0.47 | 95 | 21 | [35] |
| Zn-N-Graphene | 0.5 M $KHCO_3$ | −0.5 | 91 | 10 | [36] |
| $ZnN_4$/C | 0.5 M $KHCO_3$ | −0.43 | 95 | 4.8 | [37] |
| Sb-NC | 0.1 M $KHCO_3$ | −0.9 | 82 | 2.9 | [38] |
| Ni/Fe-N-C | 0.5 M $KHCO_3$ | −0.7 | 98 | 7.4 | [39] |
| COF-$_{366}$-Co | 0.5 M $KHCO_3$ | −0.67 | 87 | — | [40] |
| COF-$_{366}$-Co, COF-$_{367}$-Co | 0.5 M $KHCO_3$ | −0.66 | 91 | 3.3 | [41] |
| Fe porphyrin-graphene hydrogel | 0.1 M $KHCO_3$ | −0.39 | 96 | 0.42 | [42] |
| Fe(III) porphyrin/ graphene | 0.1 M $KHCO_3$ | −0.54 | 98.7 | 1.68 | [43] |

**Table 1.**
*CO-selective catalysts for the CO2RR.*

### 2.1 Metals and bimetallic materials

From both experimental and theoretical studies, Au, Ag and Zn are the most selective metals for CO formation. The $CO_2$RR on Au and Ag is characterized by low overpotentials, excellent selectivity and high activity [18–20, 30, 44]. On contrast, Zn shows relatively higher overpotentials, lower activity and moderate-to-high selectivity [8, 21].

Many bimetallic materials are demonstrated to selectively catalyze the $CO_2$RR to CO, including Cu-Sn [22–24], Cu-Zn [25], Cu-Sb [26], Cu-Ag [27], Cu-Au [28] and so on. Among all these materials, Cu-Sn catalysts have attracted the most intensive attention due to the high selectivity, good activity and outstanding repeatability. In addition, compared with others, Cu and Sn are relatively more abundant and more cost-effective, making Cu-Sn catalysts more suitable for the large-scale implementation. Hence, further study on the Cu-Sn catalysts is expected to bring benefits to both the academic and industrial sectors related to the $CO_2$ valorization.

### 2.2 Single metal atom supported on N-doped carbon

Single-metal-atom catalysts supported on porous N-doped carbon represent a class of catalysts with high atom efficiency. After introduced in 2015 by Varela et al. [30], it has gained much attention for $CO_2$ reduction. Ni supported on N-C, in contrast to Ni nanoparticles that are known to be effective in the HER, is reported to be an efficient electrocatalyst for the $CO_2$RR to CO [29, 31, 32]. Various types of Fe-N active sites have been identified and demonstrated to selectively promote the CO formation at very low overpotentials [29, 33–35]. Compared to the metallic Zn and ZnO, single atom Zn sites show much lower overpotentials where excellent CO selectivity has obtained [36, 37]. Sb atomic sites, compared to bulk Sb, $Sb_2O_3$, and Sb nanoparticles that exhibit poor activity and selectivity for the $CO_2$RR, enable the CO formation with good selectivity at relatively high overpotentials [38]. Isolated diatomic Ni-Fe sites anchored on nitrogenated carbon are also studied as an electrocatalyst for $CO_2$ reduction [39]. The catalyst exhibits high selectivity with CO Faradaic efficiency above 90% over a wide potential range from −0.5 to −0.9 V (98% at −0.7 V, vs. RHE), and robust durability.

Single atoms of selected transition metals anchored in N-doped carbon have emerged as unique and promising electrocatalysts because of the maximal atom utilization and high efficiency. Most of them perform differently from their bulk metal or oxide species, due to the metal–matrix interfacial interaction that leads to the manipulation of the electronic structures of the materials and to the emergence of additional active sites. Despite the big progress made in the recent years, many challenges remain in the development of the single atom catalysts. For example, the loading of metals is usually low, leading to relatively low geometric current density and thus limitations for practical applications. In addition, big efforts have to focus on **both** the synthetic front and structural characterizations and these necessitate the development of effective computational methods and characterization tools.

### 2.3 Immobilized molecular catalysts

Homogeneous electrocatalysis constitutes an efficient way of converting $CO_2$ to various products but some distinct challenges persist [44]. For example, the catalyst stability and recyclability are usually poor; only a small portion of the catalyst molecules at the reaction interface is active, while most of them are passive; some catalysts have poor solubility; product separation could be difficult. To overcome these disadvantages, great efforts have been dedicated to the immobilization of

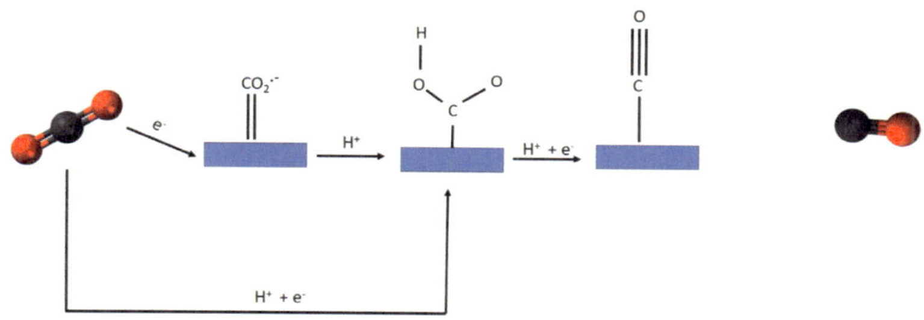


**Figure 1.**
*Schematic of possible pathways for CO production.*

molecular catalysts on electrode surfaces for the heterogeneous $CO_2$RR. Being fixed on carbon supports, the porphyrin- and phthalocyanine-based catalysts with Fe and Co centers are very selective for CO formation at relatively low overpotentials [41–43, 45]. The catalytic performance can be affected by both the intrinsic properties of the catalysts such as the structure and the metal center, and the extrinsic factors such as the catalyst immobilization methods, the support material and the catalyst loading. A deeper understanding of those intrinsic and extrinsic factors can enable the optimization of supported molecular catalysts in order to achieve the $CO_2$RR performance as high as that of the nanostructured metals, metal alloys and single atom catalysts supported on N-carbon materials [16].

The mechanism study of $CO_2$RR on metal-based materials is widely studied, in combination of in-situ spectroscopic analyses and DFT calculations [45, 46]. As shown in **Figure 1**, it is suggested that the $CO_2$RR to CO process on metallic Zn or Ag surface includes four elementary reaction steps: (1) one electron transfers to $CO_2$ to form $CO_2^{*-}$; (2) one proton transfers to $CO_2^{*-}$ to obtain $COOH^*$ intermediate; (3) an electron and a proton transfer to $COOH^*$ to form $CO^*$; (4) $CO^*$ desorbs to produce CO. Another possible pathway is supposed to include three main steps: (1) an electron coupled with a proton transfers to $CO_2$ to form $COOH^*$ intermediate; (2) another electron coupled with a proton transfers to $COOH^*$ to form $CO^*$; (3) $CO^*$ desorbs to produce CO.

## 3. Formate-selective catalysts

Due to the large storage and safety requirements for CO during carbon sequestration and storage (CCS), the production of liquid formic acid is becoming a more attractive solution. Formic acid could be directly used as a feedstock for fuel cells and as a precursor for manufacturing value-added chemicals such as formate esters, methanol, and other carboxylic acids and derivatives [47]. Some heavy metals, including Pb, Hg, In, Cd, and Tl, are efficient electrocatalysts for converting $CO_2$ into formate/formic acid. However, the defects of the high toxicity and/or high cost are standing in the way for their large-scale applications [48]. Other earth abundant metals like Sn, Cu and Bi gained a lot of attentions in recent years. **Table 2** has summarized some of the important results for formic acid production through electrocatalysis of $CO_2$.

### 3.1 Metal and metal oxides

From the pioneer work of *Hori* eta al. [57], the metals Pb, Hg, In, Sn, Cd and Tl are selective for HCOOH formation. Among them, Sn and $SnO_x$ catalysts have become the most interesting one due to the high selectivity and their non-noble, eco-friendly

| Electrocatalyst | Electrolyte | Potential (V vs. RHE) | Current Density (−mA cm$^{-2}$) | FE (%) | Reference |
|---|---|---|---|---|---|
| B-doped Pd | 0.1 M $KHCO_3$ | −0.5 V | 10 | 70 | [49] |
| S-modified Cu | 0.1 M $KHCO_3$ | −0.8 | 20 | 80 | [50] |
| Pd needles | 0.5 M $KHCO_3$ | −0.2 | 10 | 91 | [19] |
| Sn(S)/Au | 0.1 M $KHCO_3$ | −0.75 | 55 | 94 | [51] |
| nano-$SnO_2$/C | 0.1 M $NaHCO_3$ | −1.16 | 6.2 | 86.2 | [52] |
| BiOx/C | 0.5 M NaCl | −1.13 | 12.5 | 96 | [53] |
| Cu–Au | 0.5 M $KHCO_3$ | −0.6 | 10.2 | 81 | [54] |
| Bi nanotubes | 0.5 M $KHCO_3$ | −1 | 39.4 | 97 | [48] |
| Sulfur-doped indium | 0.5 M $KHCO_3$ | −0.98 | 58.9 | 95 | [55] |
| $SnO_2$ | 0.1 M $KHCO_3$ | −1.06 | 11 | 82 | [56] |

**Table 2.**
*Formate-selective catalysts for the CO2RR.*

and low-cost characteristics [56]. Recently, Pd demonstrates to be an appealing catalyst for HCOOH formation, showing high activity and good stability at extremely low overpotentials [19, 49]. Bismuth (Bi), as an HER inert metal [58], is also widely studied for the $CO_2$RR in recent years. The Bi-based catalysts are demonstrated to be a selective and active for the HCOOH production [48]. Due to the low cost and low toxicity, Bi becomes as important as Sn and is to be used in large-scale $CO_2$RR to HCOOH [59].

### 3.2 Metal sulfides

In very recent years, sulfur-modified metals have been explored as electrocatalysts, showing promising catalytic performance for the $CO_2$RR. $Cu_xS$ is one of the most intensively studied sulfides, which can selectively produce HCOOH [50]. $SnS_x$ [51], $PbS_x$ [56], $BiS_x$ [48] and $InS_x$ [55] are also demonstrated to be effective catalysts for the $CO_2$RR to HCOOH. Even though the promising performance, the role of S in the electrochemical performance is not clear until now. In order to design catalysts with higher activity, selectivity and stability, it is necessary to acquire a deeper understanding of how S functions during $CO_2$RR by performing both in-situ/operando experiments and theoretical studies.

### 3.3 Bimetallic catalysts

Compared with the pure metals, bimetallic catalysts with tuned electronic and structural properties are of particular interest. Early studies by *Hori* et al. [57] have shown that the modification of metallic surface with foreign atoms can tune the selectivity for CO or $HCOO^-$ production. Sn-based bimetallic materials are the most studied type, probably due to the high performance of Sn alone for the $CO_2$RR to HCOOH. It is worth to note that most of the Sn bimetallic materials show good HCOOH selectivity at very positive potentials, with much lower overpotentials with respect to those at the Sn/$SnO_x$ electrodes [56]. Particularly, a Cu-Au catalyst shows good selectivity and activity for the HCOOH production at even more positive potentials [54]. Until now, many studies suggested that the combination of different types of metals provide the opportunity to modulate the surface chemical

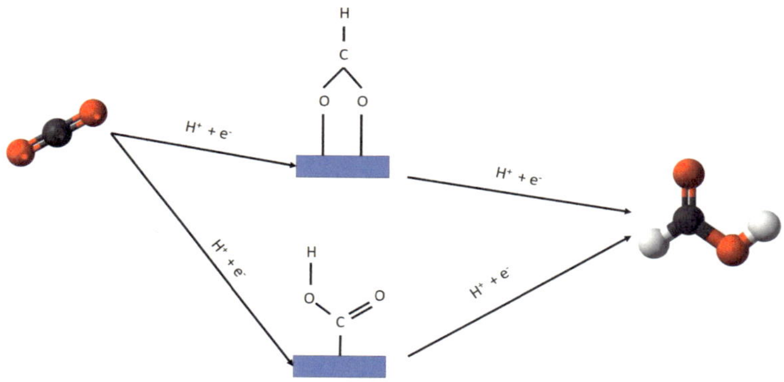


**Figure 2.**
*Schematic of possible pathways for formic acid production.*

environment and the relative binding with different intermediates, tuning the electrochemical performance of the multi-metallic catalysts in the $CO_2$RR.

In recent years, many works have been dedicated to understand the mechanism of the $CO_2$RR to HCOOH, including computational, electrokinetic and in situ analysis) [60–62]. As depicted in **Figure 2**, the formation of formate generally goes through the following pathway: 1) $CO_2\cdot^-$ radical anion is firstly formed via a one-electron transfer and bonded to the electrode surface through O atom, 2) protonation of $CO_2\cdot^-$ on the carbon atom leads to the formation of a HCOO· intermediate and 3) a second electron transfer and protonation step results in the HCOOH product [63].

## 4. $C_{1+}$ hydrocarbon selective electrocatalyst

The production of hydrocarbons through electrochemical reduction of $CO_2$ (a carbon-neutral fuel alternative to fossil fuels) is of interest because the infrastructure to store, transport and use methane and other hydrocarbons as fuel is already well established [64]. The major challenge for these products is to find the selective electrocatalysts to manage to reduce the $CO_2$ molecule with 8 and 12 electrons (methane and ethane). Considering the stability of the $CO_2$ molecule and the multi-electron-coupled-proton pathways, high energy barriers are needed to overcome for the formation of the intermediates and final product [48].

According to major reports, Cu-based materials are the main type of electrocatalysts that can produce hydrocarbon compounds including $CH_4$ and thus become the object under the most intensive study [65].

### 4.1 Cu alloys

Hirunsit et al. examined $Cu_3X$ alloys by using computational methods to examine the electrochemical reduction to $CH_4$ [66]. In an important report, Kenis and co-workers recently reported the differences between ordered, disordered, and phase-separated Cu@Pd nanoparticles with respect to product selectivity [67]. Gewirth and co-workers showed that Cu-Ag alloys from additive-controlled electrodeposition exhibited ~60% FE for $C_2H_4$ in an alkaline flow electrolyzer. In this case, by tuning the Ag-loading an optimized $C_2H_4$ selectivity can be achieved. The Ag sites were believed to play the role of a promoter for CO formation during electrochemical $CO_2$ reduction [68].

## 4.2 Other metallic alloys

Although copper gained a lot of attention for hydrocarbon production through electrochemical reduction of $CO_2$, some other metallic alloys have also shown to be quite noteworthy for this application. For example, Lewis and co-workers reported nickel–gallium alloys of different compositions prepared by drop-casting and a subsequent temperature-programmed reduction method [69]. The alloy foil was slightly enriched in zinc both at the surface and in the bulk, with a surface alloy composition of 61.3 at% zinc and a predominantly $Ag_5Zn_8$ bulk phase. The $FE_{CH4}$ values at 1.43 V vs. RHE were five times and three times higher with the alloys than those produced at pure Ag and Zn electrodes, respectively.

The summary of some recent reports for $C_{1+}$ hydrocarbons are being reported in **Table 3**.

As for the possible pathways for electrochemical reduction of $CO_2$ to hydrocarbons, In an attempt to elucidate the mechanism of $CO_2$ reduction, it was found

| Electrocatalyst | Electrolyte | Main Product | Potential (V vs. RHE) | Current Density (−mA cm$^{-2}$) | FE (%) | Reference |
|---|---|---|---|---|---|---|
| Cu nanowires/ rGO | 0.5 M $KHCO_3$ | Methane | −1.25 | 12 | 55 | [70] |
| Cu-Porphyrin | 0.5 M $KHCO_3$ | Methane | −0.98 | 15 | 47 | [71] |
| Pd decorated Cu | 0.5 M $KHCO_3$ | Methane | −0.96 | 57 | 46 | [72] |
| CuS@Ni Foam | 0.1 M $KHCO_3$ | Methane | −1.1 | 7.3 | 73 | [73] |
| Complex-derived Cu nanocluster | 0.5 M $KHCO_3$ | Methane | −1.06 | 19.7 | 66 | [74] |
| n-Cu/C | 0.1 M $NaHCO_3$ | Methane | −1.35 | 10 | 76 | [75] |
| Mesoporous Cu | 0.1 M $KHCO_3$ | Ethylene | −1.3 | 11.8 | 46 | [76] |
| $O_2$-plasma-treated Cu | 0.1 M $KHCO_3$ | Ethylene | −0.9 | 12 | 60 | [77] |
| Anodized-Cu | 0.1 M $KHCO_3$ | Ethylene | −1.08 | 19 | 38 | [78] |

**Table 3.**
*$C_1$+ hydrocarbon selective electrocatalyst.*

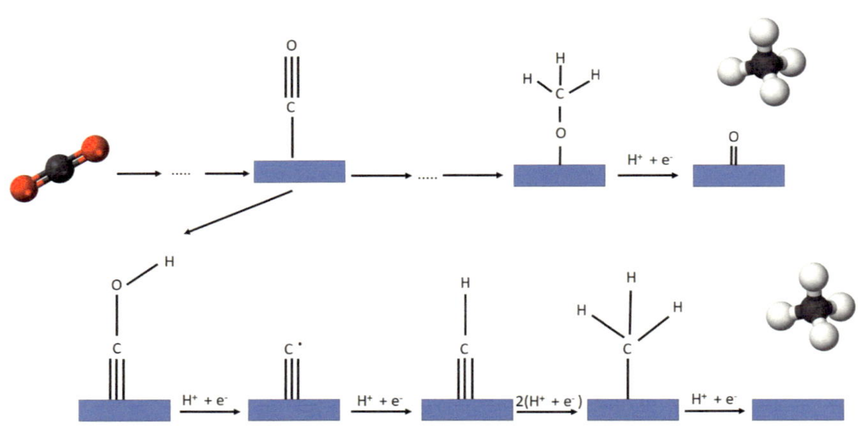


**Figure 3.**
*Schematic of possible pathways for methane production.*

that CO is a key intermediate in the formation of $CH_4$ and $C_2H_4$ [79] and that the products of $CO_2$ reduction reaction depend on the metal's binding energy to CO [80]. Based on these findings, one strategy for efficient electrochemical CO2 conversion is to separate the process into two steps: $CO_2$ reduction to CO, followed by CO reduction to oxygenates and hydrocarbons [81]. The schematic of the possible pathways toward methane production has been illustrated in **Figure 3**.

## 5. Oxygenated alcohol selective electrocatalysts

The wide range of theoretically possible products from CO to $C_{2+}$ alcohols and hydrocarbons and fuels makes the recent research to put a lot of efforts on production of more valuable products like oxygenated alcohols. The major problem as discussed before is due to a very stable structure of $CO_2$ molecule, very high activation energy needed to transform it to more attractive molecules. This high activation barrier would cause high over potentials and in case of oxygenated alcohols like methanol or ethanol high numbers of electrons (6 and 12 respectively) needed to reduce $CO_2$ molecule to desired products. So far many different metallic and alloys have been used as electrocatalysts for this application [82]. Although the performance of other product formations such as $CH_3OH$ and $C_2H_5OH$ were well below the target values, the market size of these chemicals was estimated to be much larger than those of HCOOH and CO [83]. Thus, the co-production of economically viable HCOOH and CO with other products such as $CH_4$,$C_2H_4$,$CH_3OH$, and $C_2H_5OH$ was suggested to cancel out the maximum voltage requirement [84].

### 5.1 Metal alloys

Of all metals, Cu has been identified as unique in that it is able to produce a number of "beyond CO" products such as hydrocarbons and organic oxygenates such as aldehydes and alcohols [85]. Moreover, metal alloys can adjust the binding ability of active intermediates and thus are promising to enhance the reaction selectivity and kinetics. Lu et al. [21] have synthesized an aerogel with high porosity when [BMIM][$BF_4$] and $H_2O$ with a molar ratio of 1:3 were selected as electrolytes, the faradaic efficiency (FE) and current density of $CH_3OH$ can be up to 80% and 31.8 mA/cm$^2$, respectively, over the $Pd_{83}Cu_{17}$ aerogel which attributed to the valence states, ratios, and strong interaction of Pd and Cu [86]. Also, a Zn/Ag foam electrocatalyst was prepared by Low et al. The active sites in this electrocatalyst are the strained submicron Zn dendrites, resulting in a FE of 10.5% for producing $CH_3OH$ [87].

### 5.2 Metal oxides

Metal oxide electrocatalysts have the merits of high selectivity and high energy efficiency [88]. Cuprous oxide/polypyrrole particles with octahedral and icosahedra structure ($Cu_2O$(OL-MH)/Ppy) can achieve a ultrahigh $CH_3OH$ activity and selectivity with FE of 93 ± 1.2% and 1.61 ± 0.02 μmol/(cm$^2$·s) formation rate at −0.85 V [89]. Albo and Irabien [90] used gas diffusion electrode loaded with $Cu_2O$ and achieved a FE of 42.3% for $CH_3OH$ formation, founding that Cu + can significantly affect the selectivity and activity toward $CH_3OH$. Moreover, nano $Cu_2O$ has a higher stability and selectivity compared with Cu for $CH_3OH$ production. The result of more metallic alloys and metal oxide electrocatalysts for this application have been illustrated in **Table 4**.

| Electrocatalyst | Electrolyte | Main Product | Potential (V vs. RHE) | Current Density (−mA cm$^{-2}$) | FE (%) | Reference |
|---|---|---|---|---|---|---|
| $Cu_2O/ZnO$ | 0.5 M $KHCO_3$ | Methanol | −0.7 | 6.8 | 17.7 | [91] |
| $Pd/SnO_2$ | 0.5 M$NaHCO_3$ | Methanol | −0.24 | 1.45 | 54.8 | [82] |
| Cu modified Pd | 0.5 M $KHCO_3$ | Methanol | −0.46 | 0.5 | 19.5 | [92] |
| Cu nanoparticle/ N-doped graphene | 0.1 M $KHCO_3$ | Ethanol | −1.2 | 0.7 | 63 | [93] |
| B-and-N-co-doped Nanodiamond | 0.1 M $KHCO_3$ | Ethanol | −1 | 1 | 93 | [94] |
| $Cu_2O$ films | 0.1 M $KHCO_3$ | Ethylene and Ethanol | −0.99 | 35 | 34.3 and 16.4 | [95] |

**Table 4.**
*Oxygenated alcohol selective electrocatalysts.*

It is noteworthy to mention that there are different pathways suggested for methanol and ethanol production via electrochemical reduction of $CO_2$. One possible pathway for methanol production is believed to be produced through hydrogenation of methoxy intermediate (*$OCH_3$) [44]. In detail, the *CO species is formed first. Then, the *$OCH_3$ intermediate is made from the competition between desorption of formaldehyde and the proton electron coupled transfer to formaldehyde bonded on local surface. At least, another proton electron coupled transfer occurring on *$OCH_3$ species results in methanol [65]. This possible pathway has been illustrated in **Figure 4**. In addition, the plausible pathway for ethanol

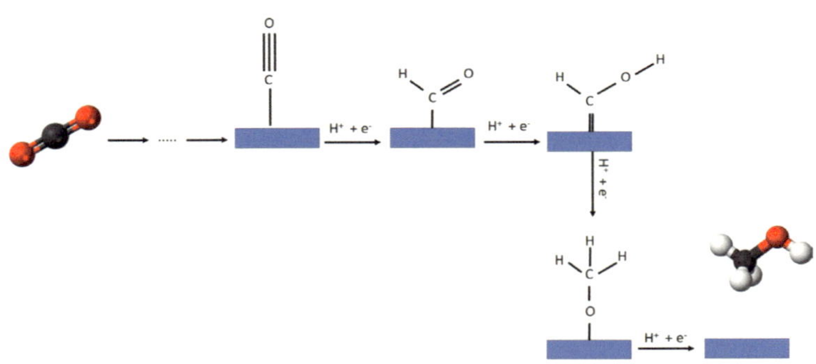

**Figure 4.**
*Schematic of possible pathways for methanol production.*

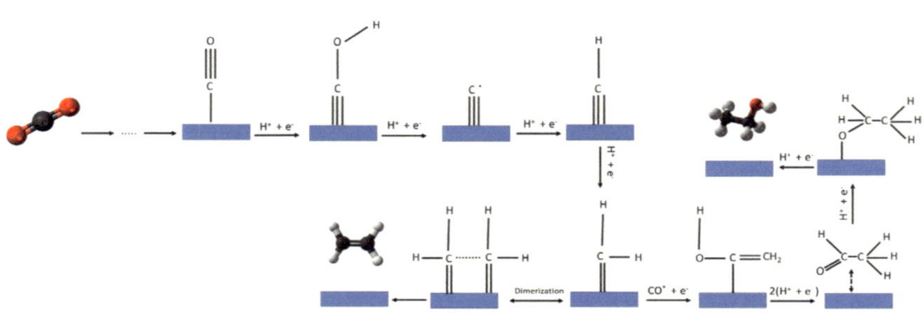


**Figure 5.**
*Schematic of possible pathways for ethylene and ethanol CO production.*

production should be discussed alongside ethylene. Ethylene is generally believed to form through either dimerization of $*CH_2$ species or proton electron coupled transfer to the carbon site of the ethylene oxide intermediate ($*OCHCH_2$) that is derived from dimerization of *CO [79]. Both routes might be the halfway leading to formation of ethanol by insertion of $^*CO$ species into $^*CH_2$ species or proton electron coupled transfer to the oxygen site of the $^*OCHCH_2$ species, correspondingly [65], as illustrated in **Figure 5**.

## 6. Conclusions

In this chapter different electrocatalysts for electrochemical reduction of $CO_2$ to value added products have been discussed. A wide range of molecules from CO and HCOOH to hydrocarbons and oxygenated alcohols are possible products of this electrochemical reaction. Up to this date the main challenge of these electrocatalytic reactions remains on scaling up and eventually industrializing the production of these value added products. The main drawback of these electrocatalytic reactions are their relatively high overpotentials and low production rate for scaling up. Although the prospective of this technology are bright, the main effort still is to find the stable, abundant electrocatalyst to be used for efficient electrocatalytic reduction of $CO_2$ at industrial scale.

## Author details

M. Amin Farkhondehfal* and Juqin Zeng
Center for Sustainable Future Technologies (CSFT@POLITO), Istituto Italiano di Tecnologia, Turin, Italy

*Address all correspondence to: amin.farkhondehfal@iit.it

## References

[1] Wang, Z.L., C.L. Li, and Y. Yamauchi, *Nanostructured nonprecious metal catalysts for electrochemical reduction of carbon dioxide.* Nano Today, 2016. **11**(3): p. 373-391.

[2] Zheng, Y., et al., *Energy related CO2 conversion and utilization: Advanced materials/nanomaterials, reaction mechanisms and technologies.* Nano Energy, 2017. **40**: p. 512-539.

[3] Kondratenko, E.V., et al., *Status and perspectives of CO2 conversion into fuels and chemicals by catalytic, photocatalytic and electrocatalytic processes.* Energy & Environmental Science, 2013. **6**(11): p. 3112-3135.

[4] Guzmán, H., et al., *How to make sustainable CO2 conversion to Methanol: Thermocatalytic versus electrocatalytic technology.* Chemical Engineering Journal, 2020: p. 127973.

[5] Francois, J.M., C. Lachaux, and N. Morin, *Synthetic Biology Applied to Carbon Conservative and Carbon Dioxide Recycling Pathways.* Frontiers in Bioengineering and Biotechnology, 2020. 7.

[6] Qiao, J., et al., *A review of catalysts for the electroreduction of carbon dioxide to produce low-carbon fuels.* Chemical Society Reviews, 2014. **43**(2): p. 631-675.

[7] Kumar, B., et al., *New trends in the development of heterogeneous catalysts for electrochemical CO2 reduction.* Catalysis Today, 2016. **270**: p. 19-30.

[8] Zeng, J., et al., *Coupled Copper–Zinc Catalysts for Electrochemical Reduction of Carbon Dioxide.* ChemSusChem, 2020. **13**(16): p. 4128-4139.

[9] Burdyny, T. and W.A. Smith, *CO2 reduction on gas-diffusion electrodes and why catalytic performance must be assessed at commercially-relevant conditions.* Energy & Environmental Science, 2019. **12**(5): p. 1442-1453.

[10] Guzmán, H., et al., *Chapter 11 - Photo/electrocatalytic hydrogen exploitation for CO2 reduction toward solar fuels production*, in *Solar Hydrogen Production*, F. Calise, et al., Editors. 2019, Academic Press. p. 365-418.

[11] Zhao, G., et al., *Progress in catalyst exploration for heterogeneous CO2 reduction and utilization: a critical review.* Journal of Materials Chemistry A, 2017. **5**(41): p. 21625-21649.

[12] Xie, H., et al., *Cu-based nanocatalysts for electrochemical reduction of CO2.* Nano Today, 2018. **21**: p. 41-54.

[13] Hori, Y., K. Kikuchi, and S. Suzuki, *Production of Co and Ch4 in Electrochemical Reduction of Co2 at Metal-Electrodes in Aqueous Hydrogencarbonate Solution.* Chemistry Letters, 1985(11): p. 1695-1698.

[14] Yin, Z., G.T.R. Palmore, and S. Sun, *Electrochemical Reduction of CO2 Catalyzed by Metal Nanocatalysts.* Trends in Chemistry, 2019. **1**(8): p. 739-750.

[15] Hernandez, S., et al., *Syngas production from electrochemical reduction of CO2: current status and prospective implementation.* Green Chemistry, 2017. **19**(10): p. 2326-2346.

[16] Nielsen, D.U., et al., *Chemically and electrochemically catalysed conversion of CO2 to CO with follow-up utilization to value-added chemicals.* Nature Catalysis, 2018. **1**(4): p. 244-254.

[17] Verma, S., et al., *A Gross-Margin Model for Defining Technoeconomic Benchmarks in the Electroreduction of CO2.* ChemSusChem, 2016. **9**(15): p. 1972-1979.

[18] Zhu, W., et al., *Monodisperse Au Nanoparticles for Selective Electrocatalytic Reduction of CO2 to CO.* Journal of the American Chemical Society, 2013. **135**(45): p. 16833-16836.

[19] Liu, M., et al., *Enhanced electrocatalytic CO2 reduction via field-induced reagent concentration.* Nature, 2016. **537**(7620): p. 382-386.

[20] Xi, W., et al., *Ultrathin Ag Nanowires Electrode for Electrochemical Syngas Production from Carbon Dioxide.* ACS Sustainable Chemistry & Engineering, 2018. **6**(6): p. 7687-7694.

[21] Lu, Q., et al., *A selective and efficient electrocatalyst for carbon dioxide reduction.* Nature Communications, 2014. **5**(1): p. 3242.

[22] Feng, Y., et al., *Electroreduction of Carbon Dioxide in Metallic Nanopores through a Pincer Mechanism.* Angew Chem Int Ed Engl, 2020. **59**(43): p. 19297-19303.

[23] Zeng, J., et al., *Advanced Cu-Sn foam for selectively converting CO2 to CO in aqueous solution.* Applied Catalysis B: Environmental, 2018. **236**: p. 475-482.

[24] Sarfraz, S., et al., *Cu–Sn Bimetallic Catalyst for Selective Aqueous Electroreduction of CO2 to CO.* ACS Catalysis, 2016. **6**(5): p. 2842-2851.

[25] Qing Li, J.F., Wenlei Zhu, Zhengzheng Chen, Bo Shen, Liheng Wu,Zheng Xi,Tanyuan Wang, Gang Lu, Jun-jie Zhu,Shouheng Sun, *Tuning Sn-Catalysis for Electrochemical Reduction of CO2 to CO via the Core/Shell Cu/SnO2 Structure.* J. Am. Chem. Soc, 2017. **139**(12): p. 4290-4293.

[26] Moreno-García, P., et al., *Selective Electrochemical Reduction of CO2 to CO on Zn-Based Foams Produced by Cu2+ and Template-Assisted Electrodeposition.* ACS Applied Materials & Interfaces, 2018. **10**(37): p. 31355-31365.

[27] Li, Y., et al., *Achieving Highly Selective Electrocatalytic CO2 Reduction by Tuning CuO-Sb2O3 Nanocomposites.* ACS Sustainable Chemistry & Engineering, 2020. **8**(12): p. 4948-4954.

[28] Cai, Z., et al., *Unlocking Bifunctional Electrocatalytic Activity for CO2 Reduction Reaction by Win-Win Metal–Oxide Cooperation.* ACS Energy Letters, 2018. **3**(11): p. 2816-2822.

[29] Kim, D., et al., *Synergistic geometric and electronic effects for electrochemical reduction of carbon dioxide using gold–copper bimetallic nanoparticles.* Nature Communications, 2014. **5**(1): p. 4948.

[30] Varela, A.S., et al., *Metal-Doped Nitrogenated Carbon as an Efficient Catalyst for Direct CO2 Electroreduction to CO and Hydrocarbons.* Angewandte Chemie International Edition, 2015. **54**(37): p. 10758-10762.

[31] Koshy, D.M., et al., *Understanding the Origin of Highly Selective CO2 Electroreduction to CO on Ni,N-doped Carbon Catalysts.* Angewandte Chemie International Edition, 2020. **59**(10): p. 4043-4050.

[32] Möller, T., et al., *Efficient CO2 to CO electrolysis on solid Ni–N–C catalysts at industrial current densities.* Energy & Environmental Science, 2019. **12**(2): p. 640-647.

[33] Li, X., et al., *Isolated FeN(4) Sites for Efficient Electrocatalytic CO(2) Reduction.* Adv Sci (Weinh), 2020. **7**(17): p. 2001545.

[34] Zhang, H., et al., *A Graphene-Supported Single-Atom FeN5 Catalytic Site for Efficient Electrochemical CO2 Reduction.* Angewandte Chemie International Edition, 2019. **58**(42): p. 14871-14876.

[35] Gu, J., et al., *Atomically dispersed Fe3+ sites catalyze efficient CO2 electroreduction to CO.* Science, 2019. **364**(6445): p. 1091-1094.

[36] Chen, Z., et al., *Zinc-Coordinated Nitrogen-Codoped Graphene as an Efficient Catalyst for Selective Electrochemical Reduction of CO2 to CO.* ChemSusChem, 2018. **11**(17): p. 2944-2952.

[37] Yang, F., et al., *Highly Efficient CO2 Electroreduction on ZnN4-based Single-Atom Catalyst.* Angewandte Chemie International Edition, 2018. **57**(38): p. 12303-12307.

[38] Jia, M., et al., *Single Sb sites for efficient electrochemical CO2 reduction.* Chemical Communications, 2019. **55**(80): p. 12024-12027.

[39] Ren, W., et al., *Isolated Diatomic Ni-Fe Metal–Nitrogen Sites for Synergistic Electroreduction of CO2.* Angewandte Chemie International Edition, 2019. **58**(21): p. 6972-6976.

[40] Diercks, C.S., et al., *Reticular Electronic Tuning of Porphyrin Active Sites in Covalent Organic Frameworks for Electrocatalytic Carbon Dioxide Reduction.* Journal of the American Chemical Society, 2018. **140**(3): p. 1116-1122.

[41] Lin, S., et al., *Covalent organic frameworks comprising cobalt porphyrins for catalytic CO2 reduction in water.* Science, 2015. **349**(6253): p. 1208-1213.

[42] Choi, J., et al., *Energy efficient electrochemical reduction of CO2 to CO using a three-dimensional porphyrin/graphene hydrogel.* Energy & Environmental Science, 2019. **12**(2): p. 747-755.

[43] Choi, J., et al., *A Porphyrin/Graphene Framework: A Highly Efficient and Robust Electrocatalyst for Carbon Dioxide Reduction.* Advanced Energy Materials, 2018. **8**(26): p. 1801280.

[44] Peterson, A.A., et al., *How copper catalyzes the electroreduction of carbon dioxide into hydrocarbon fuels.* Energy & Environmental Science, 2010. **3**(9): p. 1311-1315.

[45] Qin, B., et al., *Electrochemical Reduction of CO2 into Tunable Syngas Production by Regulating the Crystal Facets of Earth-Abundant Zn Catalyst.* ACS Applied Materials & Interfaces, 2018. **10**(24): p. 20530-20539.

[46] Rosen, J., et al., *Mechanistic Insights into the Electrochemical Reduction of CO2 to CO on Nanostructured Ag Surfaces.* ACS Catalysis, 2015. **5**(7): p. 4293-4299.

[47] Ren, M., et al., *CO2 to Formic Acid Using Cu–Sn on Laser-Induced Graphene.* ACS Applied Materials & Interfaces, 2020. **12**(37): p. 41223-41229.

[48] Fan, K., et al., *Curved Surface Boosts Electrochemical CO2 Reduction to Formate via Bismuth Nanotubes in a Wide Potential Window.* ACS Catalysis, 2020. **10**(1): p. 358-364.

[49] Jiang, B., et al., *Boosting Formate Production in Electrocatalytic CO(2) Reduction over Wide Potential Window on Pd Surfaces.* J Am Chem Soc, 2018. **140**(8): p. 2880-2889.

[50] Shinagawa, T., et al., *Sulfur-Modified Copper Catalysts for the Electrochemical Reduction of Carbon Dioxide to Formate.* ACS Catalysis, 2018. **8**(2): p. 837-844.

[51] Zheng, X., et al., *Sulfur-Modulated Tin Sites Enable Highly Selective Electrochemical Reduction of CO2 to Formate.* Joule, 2017. **1**(4): p. 794-805.

[52] Zhang, S., P. Kang, and T.J. Meyer, *Nanostructured Tin Catalysts for Selective Electrochemical Reduction of Carbon Dioxide to Formate.* Journal of the American Chemical Society, 2014. **136**(5): p. 1734-1737.

[53] Lee, C.W., et al., *Selective Electrochemical Production of Formate from Carbon Dioxide with Bismuth-Based Catalysts in an Aqueous Electrolyte.* ACS Catalysis, 2018. **8**(2): p. 931-937.

[54] Tao, Z., et al., *Copper–Gold Interactions Enhancing Formate Production from Electrochemical CO2 Reduction.* ACS Catalysis, 2019. **9**(12): p. 10894-10898.

[55] Ma, W., et al., *Promoting electrocatalytic CO2 reduction to formate via sulfur-boosting water activation on indium surfaces.* Nature Communications, 2019. **10**(1): p. 892.

[56] Bejtka, K., et al., *Chainlike Mesoporous SnO2 as a Well-Performing Catalyst for Electrochemical CO2 Reduction.* ACS Applied Energy Materials, 2019. **2**(5): p. 3081-3091.

[57] Hori, Y., et al., *Electrocatalytic process of CO selectivity in electrochemical reduction of CO2 at metal electrodes in aqueous media.* Electrochimica Acta, 1994. **39**(11): p. 1833-1839.

[58] Greeley, J., et al., *Computational high-throughput screening of electrocatalytic materials for hydrogen evolution.* Nature Materials, 2006. **5**(11): p. 909-913.

[59] Díaz-Sainz, G., et al., *CO2 electroreduction to formate: Continuous single-pass operation in a filter-press reactor at high current densities using Bi gas diffusion electrodes.* Journal of CO2 Utilization, 2019. **34**: p. 12-19.

[60] Yoo, J.S., et al., *Theoretical Insight into the Trends that Guide the Electrochemical Reduction of Carbon Dioxide to Formic Acid.* ChemSusChem, 2016. **9**(4): p. 358-363.

[61] Luc, W., et al., *Ag–Sn Bimetallic Catalyst with a Core–Shell Structure for CO2 Reduction.* Journal of the American Chemical Society, 2017. **139**(5): p. 1885-1893.

[62] Li, C.W. and M.W. Kanan, *CO2 Reduction at Low Overpotential on Cu Electrodes Resulting from the Reduction of Thick Cu2O Films.* Journal of the American Chemical Society, 2012. **134**(17): p. 7231-7234.

[63] Zhao, S., et al., *Advances in Sn-Based Catalysts for Electrochemical CO2 Reduction.* Nano-Micro Letters, 2019. **11**(1): p. 62.

[64] Vogt, C., et al., *The renaissance of the Sabatier reaction and its applications on Earth and in space.* Nature Catalysis, 2019. **2**(3): p. 188-197.

[65] Long, C., et al., *Electrochemical Reduction of CO2 over Heterogeneous Catalysts in Aqueous Solution: Recent Progress and Perspectives.* Small Methods, 2019. **3**(3): p. 1800369.

[66] Hirunsit, P., W. Soodsawang, and J. Limtrakul, *CO2 Electrochemical Reduction to Methane and Methanol on Copper-Based Alloys: Theoretical Insight.* The Journal of Physical Chemistry C, 2015. **119**(15): p. 8238-8249.

[67] Ma, S., et al., *Electroreduction of Carbon Dioxide to Hydrocarbons Using Bimetallic Cu–Pd Catalysts with Different Mixing Patterns.* Journal of the American Chemical Society, 2017. **139**(1): p. 47-50.

[68] Hoang, T.T.H., et al., *Nanoporous Copper–Silver Alloys by Additive-Controlled Electrodeposition for the Selective Electroreduction of CO2 to Ethylene and Ethanol.* Journal of the American Chemical Society, 2018. **140**(17): p. 5791-5797.

[69] Torelli, D.A., et al., *Nickel–Gallium-Catalyzed Electrochemical Reduction of CO2 to Highly Reduced Products at Low Overpotentials.* ACS Catalysis, 2016. **6**(3): p. 2100-2104.

[70] Li, Y., et al., *Structure-Sensitive CO(2) Electroreduction to Hydrocarbons on Ultrathin 5-fold Twinned Copper Nanowires.* Nano Lett, 2017. **17**(2): p. 1312-1317.

[71] Weng, Z., et al., *Electrochemical CO2 Reduction to Hydrocarbons on a Heterogeneous Molecular Cu Catalyst in Aqueous Solution.* J Am Chem Soc, 2016. **138**(26): p. 8076-9.

[72] Weng, Z., et al., *Self-Cleaning Catalyst Electrodes for Stabilized CO(2) Reduction to Hydrocarbons.* Angew Chem Int Ed Engl, 2017. **56**(42): p. 13135-13139.

[73] Zhao, Z., et al., *Efficient and stable electroreduction of CO2 to CH4 on CuS nanosheet arrays.* Journal of Materials Chemistry A, 2017. **5**(38): p. 20239-20243.

[74] Weng, Z., et al., *Active sites of copper-complex catalytic materials for electrochemical carbon dioxide reduction.* Nat Commun, 2018. **9**(1): p. 415.

[75] Manthiram, K., B.J. Beberwyck, and A.P. Alivisatos, *Enhanced Electrochemical Methanation of Carbon Dioxide with a Dispersible Nanoscale Copper Catalyst.* Journal of the American Chemical Society, 2014. **136**(38): p. 13319-13325.

[76] Yang, K.D., et al., *Morphology-Directed Selective Production of Ethylene or Ethane from CO(2) on a Cu Mesopore Electrode.* Angew Chem Int Ed Engl, 2017. **56**(3): p. 796-800.

[77] Mistry, H., et al., *Highly selective plasma-activated copper catalysts for carbon dioxide reduction to ethylene.* Nat Commun, 2016. **7**: p. 12123.

[78] Lee, S.Y., et al., *Mixed Copper States in Anodized Cu Electrocatalyst for Stable and Selective Ethylene Production from CO(2) Reduction.* J Am Chem Soc, 2018. **140**(28): p. 8681-8689.

[79] Schouten, K.J.P., et al., *Two Pathways for the Formation of Ethylene in CO Reduction on Single-Crystal Copper Electrodes.* Journal of the American Chemical Society, 2012. **134**(24): p. 9864-9867.

[80] Kuhl, K.P., et al., *New insights into the electrochemical reduction of carbon dioxide on metallic copper surfaces.* Energy & Environmental Science, 2012. **5**(5): p. 7050-7059.

[81] Verdaguer-Casadevall, A., et al., *Probing the Active Surface Sites for CO Reduction on Oxide-Derived Copper Electrocatalysts.* Journal of the American Chemical Society, 2015. **137**(31): p. 9808-9811.

[82] Zhang, W., et al., *Electrochemical Reduction of Carbon Dioxide to Methanol on Hierarchical Pd/SnO2 Nanosheets with Abundant Pd–O–Sn Interfaces.* Angewandte Chemie International Edition, 2018. **57**(30): p. 9475-9479.

[83] Jouny, M., W. Luc, and F. Jiao, *General Techno-Economic Analysis of CO2 Electrolysis Systems.* Industrial & Engineering Chemistry Research, 2018. **57**(6): p. 2165-2177.

[84] Lee, M.-Y., et al., *Current achievements and the future direction of electrochemical CO2 reduction: A short review.* Critical Reviews in Environmental Science and Technology, 2020. **50**(8): p. 769-815.

[85] Kim, C., et al., *Alloy Nanocatalysts for the Electrochemical Oxygen Reduction (ORR) and the Direct Electrochemical Carbon Dioxide Reduction Reaction (CO2RR).* Advanced Materials, 2019. **31**(31): p. 1805617.

[86] Lu, L., et al., *Highly Efficient Electroreduction of CO2 to Methanol on Palladium–Copper Bimetallic Aerogels.* Angewandte Chemie International Edition, 2018. **57**(43): p. 14149-14153.

[87] Low, Q.H., et al., *Enhanced Electroreduction of Carbon Dioxide to*

*Methanol Using Zinc Dendrites Pulse-Deposited on Silver Foam.* Angewandte Chemie International Edition, 2019. **58**(8): p. 2256-2260.

[88] Liu, Y., et al., *Recent progress on electrochemical reduction of CO2 to methanol.* Current Opinion in Green and Sustainable Chemistry, 2020. **23**: p. 10-17.

[89] Periasamy, A.P., et al., *Facet- and structure-dependent catalytic activity of cuprous oxide/polypyrrole particles towards the efficient reduction of carbon dioxide to methanol.* Nanoscale, 2018. **10**(25): p. 11869-11880.

[90] Hazarika, J. and M.S. Manna, *Electrochemical reduction of CO2 to methanol with synthesized Cu2O nanocatalyst: Study of the selectivity.* Electrochimica Acta, 2019. **328**: p. 135053.

[91] Albo, J., et al., *Production of methanol from CO2 electroreduction at Cu2O and Cu2O/ZnO-based electrodes in aqueous solution.* Applied Catalysis B: Environmental, 2015. **176-177**: p. 709-717.

[92] Zhang, F.-Y., et al., *Cu overlayers on tetrahexahedral Pd nanocrystals with high-index facets for CO<sub>2</sub> electroreduction to alcohols.* Chemical communications (Cambridge, England), 2017. **53**(57): p. 8085-8088.

[93] Song, Y., et al., *High-Selectivity Electrochemical Conversion of CO2 to Ethanol using a Copper Nanoparticle/N-Doped Graphene Electrode.* ChemistrySelect, 2016. **1**(19): p. 6055-6061.

[94] Liu, Y., et al., *Selective Electrochemical Reduction of Carbon Dioxide to Ethanol on a Boron- and Nitrogen-Co-doped Nanodiamond.* Angewandte Chemie (International ed. in English), 2017. **56**(49): p. 15607-15611.

[95] Ren, D., et al., *Selective Electrochemical Reduction of Carbon Dioxide to Ethylene and Ethanol on Copper(I) Oxide Catalysts.* ACS Catalysis, 2015. **5**(5): p. 2814-2821.

Chapter 5

# Investigation of Zn/Ni-Based Electrocatalysts for Electrochemical Conversion of $CO_2$ to SYNGAS

*Mohammadali Beheshti, Saeid Kakooei, Mokhtar Che Ismail and Shohreh Shahrestani*

## Abstract

In the last decade, there is some research on the conversion of $CO_2$ to energy form. $CO_2$ can be converted to value-added chemicals including HCOOH, CO, $CH_4$, $C_2H_4$, and liquid hydrocarbons that can be used in various industries. Among the methods, electrochemical methods are of concern regarding their capability to operate with an acceptable reaction rate and great efficiency at room temperature and can be easily coupled with renewable energy sources. Besides, electrochemical cell devices have been manufactured in a variety of sizes, from portable to large-scale applications. Catalysts that optionally reduce $CO_2$ at low potential are required. Therefore, choosing a suitable electrocatalyst is very important. This chapter focused on the electrochemical reduction of $CO_2$ by Zn-Ni bimetallic electrocatalyst. The Zn-Ni coatings were deposited on the low-carbon steel substrate. Electrochemical deposition parameters such as temperature in terms of LPR corrosion rate, microstructure, microcracks, and its composition have been investigated. Then, the electrocatalyst stability and activity, as well as gas intensity and selectivity, were inspected by SEM/EDX analysis, GC, and electrochemical tests. Among the electrocatalysts for $CO_2$ reduction reaction, the $Zn_{65\%}$-$Ni_{35\%}$ electrode with cluster-like microstructure had the best performance for $CO_2$ reduction reaction according to minimum coke formation (<10%) and optimum CO and $H_2$ faradaic efficiencies (CO FE% = 55% and H2 FE% = 45%).

**Keywords:** electrocatalyst, electrochemical method, $CO_2$ reduction reaction, Zn-Ni, energy conversion, pollution, catalyst activity and stability

## 1. Introduction

Carbon dioxide is a chemical compound made up of one carbon atom and two oxygen atoms. It is existing in minimal concentrations in the atmosphere and behaves as a greenhouse gas that promotes environmental warming and pollution. However, carbon dioxide can be used as a source of high-value chemicals, as a source of sustainable energy. So far, many activities have been done to convert $CO_2$ into chemical materials, which can be applied as fuel for the industries.

With the increasing demand for energy and population growth, $CO_2$ emissions have grown as a by-product of power and industrial plants. In the last decade, $CO_2$ conversion has increased to other beneficial products. This process is useful for reducing pollution and warming of the earth. Developing a variety of electrocatalysts with high efficiency and good stability is a crucial issue [1].

The electrochemical $CO_2$ reaction reduction in recent decades has become crucial because it is a good reaction to artificial fuels and energy storage. When this process is linked to renewable energy sources such as solar cells, it can be a good alternative to fossil fuels. It also reduces $CO_2$ emissions in the atmosphere. But there are major problems for the reaction of $CO_2$ reduction, which includes low efficiency and low catalytic activity with cost-effective catalysts. Therefore, there is an important challenge in the present research, so that catalyst with better selectivity and higher activity and stability can be developed [2].

In recent years, several studies were done on various electrocatalysts, but yet, there are problems in Faradaic Efficiency (FE), Current Density (CD), Energy Efficiency, electrocatalyst deactivate, the internal resistance of electrocatalysts, and the potential for scalability to the large sizes without the loss of efficiency, because $CO_2$ is a thermodynamically stable molecule, it is fully oxidized [3–12]. A suitable electrocatalyst to reduce $CO_2$ is necessary to reach a low-cost process with acceptable selectivity and efficiency. In recent decades, the electrochemical reduction of $CO_2$ has interested a lot of consideration as low-cost electricity can come from renewable sources of energy such as solar and wind [13–18].

### 1.1 Zn-Ni coating

Zinc as another choice of cadmium has been studied for its ability to resist corrosion regarding its sacrificial properties and has demonstrated its ability to provide adequate corrosion behavior results through the study conducted on mechanical properties and corrosion protection of Zn electrodeposition [19]. Though Zn is considered a possible option, its corrosion behavior does not look acceptable in an aggressive condition with greater temperatures. Electrodeposited Zn coatings study tests indicated that pure Zn has weak corrosion resistance properties compared to cadmium [20]. Therefore, the need for metal coatings with corrosion properties outstanding to those of pure Zn and comparable or improved to cadmium has driven the industrial production of electrodeposits involving Zn alloys with VIIIB-group metals (e.g. Zn-Fe, Zn-Ni, Zn-Co) [21]. The electrodeposition of Zn and its eight-group metals including Co, Fe, and Ni have been extensively investigated and analyzed for their ability to be an excellent corrosion resistant alloy.

#### *1.1.1 Zinc-nickel alloy corrosion behavior*

In recent years, a lot of research has been performed to investigate the possibility that the Zn-Ni alloy could be a substitute with a corrosion property corresponding to the toxic coatings of cadmium. Much research has also been done to distinguish and determine the corrosion resistance of the Zn-Ni coatings [22–25]. The corrosion resistance of deposited Zn-Ni coatings on steel substrate indicated as having the acceptable corrosion property (corrosion rate: ~ 11 mm/year) was reached for Zn-Ni alloys in the range from 12 to 15 wt.% of Ni content in the coating so that the coating with Ni content from 12 to 15 wt.% maintains the anodic behavior of the steel, retaining the sacrificial behavior with a decrease corrosion rate after the addition of Ni, which increases the potential nearer to the substrate providing protection for a too time [21]. This has been endorsed by reports conducted by other authors [22–25] who have stated that Zn-Ni coating with a Ni amount of 12 to 15 wt.% supplies adequate corrosion protection. While the coating retains its sacrificial behavior regarding the steel substrate, whenever the alloy with more than 30 wt.% of Ni turns nobler than the substrate, missing its sacrificial behavior. Hence, it led to preferential corrosion of the steel, and Ni amount of less than 10 wt.% in the coating produced smaller barrier performance. Byk et al. [25] performed tests showing the greatest corrosion resistance properties utilizing a poor acid chloride solution with the Zn-(15 wt.%) Ni coating having the least corrosion CD, demonstrating the best corrosion protection, and this is qualified to the existence of the γ phase ($Ni_5Zn_{21}$) which is gained with Zn-Ni coatings with Ni amount from 12 to 15 wt.% [25]. The coatings of Zn-Ni coating with 10–15 wt.% of Ni have more suitable corrosion resistance, better weldability, and superior formability. The presence of Ni in the Zn-Ni alloy in the optimal range from 12 to 15 wt.% reduces the rate of Zn dissolution, supplying greater and longer corrosion resistance than pure Zn [24].

### 1.2 $CO_2$ reduction reaction

Environments change due to greenhouse gases ($CO_2$) is a significant hazard to the protection of human society. The capture and conversion of carbon to the value-added chemical are attended to be the most agreeable method to prevent the rise of $CO_2$ in the environment as seen in **Figure 1**. But the cost of high technology accessible to capture, store, and convert $CO_2$ stops its functional operation [26]. Recycling $CO_2$ and transforming it into value-added chemicals create challenges for researchers in the area of catalysts. Among the various methods, the electrochemical method has unique advantages [27–29]. Most of the electrochemical reactions can be seen in small to industrial conditions. Besides, if the electricity is required from renewable sources, these sources of energy generate the required electricity, $CO_2$ will not be produced and, therefore, will have a good effect on the worldwide $CO_2$ level [30].

Studies showed that CO was an intermediator and also methane ($CH_4$) or ethylene ($C_2H_4$) was generated from HCO* or COH* intermediates. Norskov et al. presented details of reaction pathways to produce $C_2H_4$ and $CH_4$ from the $CO_2$ reduction reaction at copper catalysts using the Density Functional Theory (DFT) [31–34].

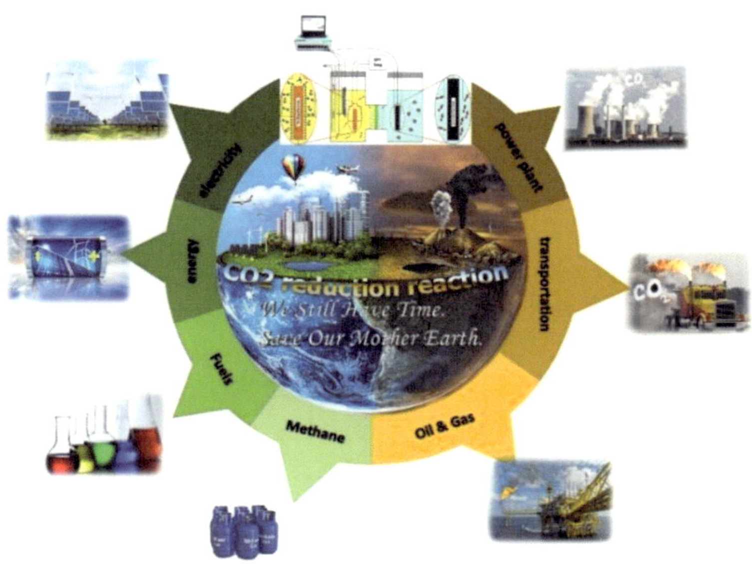


**Figure 1.**
*Graphical schematic of the influence of $CO_2$ on the globe and conversion of $CO_2$ to the useable energy.*

The outcomes demonstrated that the formation of HCO* was a key step for the reaction. They also compared the carbon dioxide reduction reaction in several transition-metal electrodes and determined that copper is the most efficient electrode for this case [35]. In the electrolysis of $CO_2$, the anode and cathode were located on separate sides, which were interconnected with a membrane in the middle of them. In the anode, the water oxidized to ion hydrogen ($H^+$) and molecular oxygen ($O_2$), while in the cathode, $CO_2$ was reduced to carbon compounds, and hydrogen was reduced [36]. The electrocatalysts for the reduction reaction of $CO_2$ totally divided into a few different classes as seen in **Figure 2**. Metals such as Ni, Pt, Al, Fe, Ti, and Ga were used as the catalysts for $H_2$ production, and CO was not created as the main product [50–53]. The $H_2$ evolution reaction rates by these group metals are commonly greater than that of the $CO_2$RR rate.

Another class of metals of Ag, Au, and Zn convert $CO_2$ to CO with an acceptable efficiency [54]. Catalysts consist of In, Pb, Hg, and Sn convert $CO_2$ to formate as the main product. On these metals, the mechanism of $CO_2$RR to formate is different in which there is no breaking of the C-O bond. Electrodes including W, Cr, and Mo have been reported as inadequate catalysts because of weak selectivity and reduction rate. Copper as a metal catalyst can react to a reduction in $CO_2$ to alcohol and hydrocarbons ($C_2H_4$, $CH_4$, $CH_3OH$). However, recent research had shown that the $CO_2$RR to these fuels was made at lower efficiency, which was influenced by the binding-energy of the intermediate species of CO. For example, Ag and Au catalysts can produce CO more rather due to less energy for intermediate carbon monoxide molecules. Since it can be evolved from the surface without more reaction. Therefore, producing higher carbon species at these levels is extremely minimal. However, Cu is a unique catalyst that allows it to produce various carbonaceous products (such as, alcohol and hydrocarbons) with higher activity [54].

Electrodes play a key role in all reactions according to heterogeneous electrochemical reactions, such as $CO_2$RR [55]. The durability and performance of the electrochemical cells are essentially defined by the processes happening at the electrolyte-electrode interface. Overall, electrodes include an electrocatalyst layer as well as a backing layer or substrate that attend multiple acts: firstly, to transport reactant gases, $CO_2$, from the electrolyte to the catalyst layer; secondly, to derive products from the catalyst layer into the membrane/electrolyte; and lastly, electrons connectivity with little resistance [55–58]. Most electrode efficiency, and accordingly electrochemical cell efficiency, requires enhancing all these three processes that greatly relate to the complicated microstructure of the electrodes. Till now, the

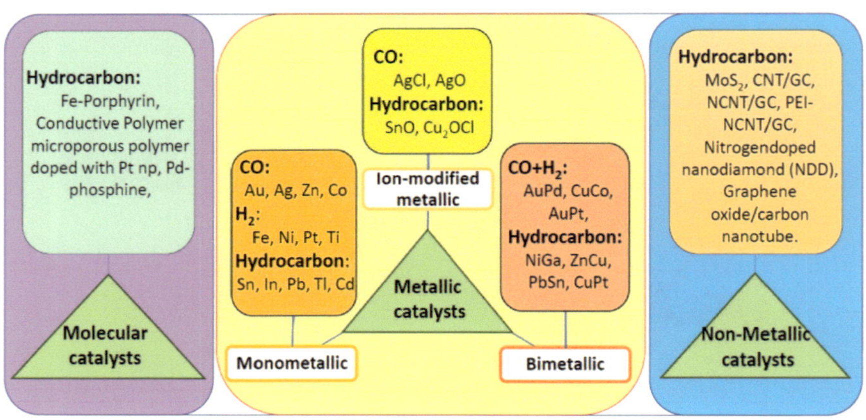


**Figure 2.**
*A set of three main categories of electrocatalysts for $CO_2$ reduction reaction [36, 37–49].*

nanoparticles of Ag [59, 60], Sn [61], Au [62], $MoO_2$ [63, 64], Bi [65], $MoS_2$ [62], etc., coated on low carbon steel substrate and $Cu_2O/TiO_2$/FTO [66] have been utilized to convert $CO_2$ to CO applying room-temperature ionic liquids (RTILs) as electrocatalysts. Nevertheless, none of these materials enabled the development of CO with a CD of >100 mA/cm$^2$ in $CO_2$RR during controlled potential electrolysis (CPE) tests in combination with any of the utilized RTIL assistant catalysts, which is required to commercially use any of these procedures. In the last decade, the electrochemical $CO_2$RR had been widely considered [67–69]. The reduction reaction products of electrochemical $CO_2$RR on the Cu-based electrodes are hydrocarbons for example $C_2H_4$ and $CH_4$ [70–72]. Practical investigations on the electrochemical $CO_2$RR in the base electrodes of copper showed that the exhaust gas contains CO, $CH_4$, $C_2H_4$, and primary alcohol that depended on their electrolyte [73, 74]. There were numerous studies of electrochemically $CO_2$ reduction reaction on Cu-based electrodes [37–39].

**Table 1** shows the summarized characterization of electrocatalysts for the $CO_2$RR to various. As shown in **Table 1** and **Figure 3** for SYNGAS (CO + $H_2$) production,

| Electrocatalysts | Faradaic Efficiency | Current Density (mA/cm$^2$) | Main product | Ref. |
|---|---|---|---|---|
| Fe-Porphyrin | — | < 5 | **Hydrocarbon** | [75] |
| Sn-based | 91% | 15 | | [61] |
| nano-$SnO_2$ | 80% | 9 | | [76] |
| Sn foil | <20% | < 5 | | [76] |
| $Sn/SnO_x$ nanoclusters | <40% | — | | [76] |
| Ni-Ga | <70% | <1 | | [77] |
| Molybdenum disulphide | ~90% | 65 | | [62] |
| Cl-induced bi-phasic Cu2O–Cu | ~70% | <0.2 | | [76] |
| Conductive Polymer microporous polymer doped with Pt np | >95% | — | | [78] |
| Strontium-doped lanthanum | <15% | <20 | | [79] |
| Iridium/Ruthenium Oxide | ~90% | | | [80] |
| Zn-Cu | 60% | 4 | | [49] |
| Pb - Sn | — | 2.5 | | [35] |
| Zn-Co based | 70%- | 8 | | [3] |
| Cu-Pt | <38% | — | | [81] |
| Cu-Ni | <40% | — | | [82] |
| Ni-based | 80% | 5 | | [3] |
| CNT/GC | <10% | ~2 | | [76] |
| NCNT/GC | 60% | ~4 | | [76] |
| PEI-NCNT/GC | <85% | ~10 | | [76] |
| Nitrogendoped nanodiamond | 90% | ~1 | | [76] |
| Pd –based | <50% | <5 | **(CO + $H_2$) Syngas** | [83] |
| $Au_{0.76}$ –$Pd_{0.24}$ | ~90% | <10 | | [84] |
| Cu-Co | 33% | 8 | | [85] |
| $Au_{0.78}$ –$Pt_{0.22}$ | ~90% | 0.6 | | [86] |
| Cu-Ag | ~64% | 10 | | [87] |

| Electrocatalysts | Faradaic Efficiency | Current Density ($mA/cm^2$) | Main product | Ref. |
|---|---|---|---|---|
| Bismuth- based | <90% | <4 | Carbon monoxide (CO) | [65] |
| [EMIM]$BF_4$-Bi | 93% | 4 | | [76] |
| Graphene oxide/carbon nanotube | 85% | 2.3 | | [88] |
| carbon paper | 68% | 0.9 | | [89] |
| Sn/$SnO_x$ nanoclusters | <60% | — | | [76] |
| Zn electrode | 64% | 2.8 | | [76] |
| Zn porous | 77.9% | 8 | | [76] |
| Zn dendrite | 80% | 12 | | [49] |
| Zn hexagonal | ~83% | 4.4 | | [75] |
| Cu polycrystalline | 30% | — | | [76] |
| Au np | >90% | 81 | | [43] |
| Au bulk | 80% | 2.2 | | [45] |
| Oxide Derived-Au | — | 10 | | [40] |
| Ag bulk | 82% | 2.0 | | [59] |
| Ag np | ~80% | 29 | | [44] |
| Ag np | 92% | 10 | | [46] |

**Table 1.**
*Product distribution for electrochemical $CO_2$ reduction reaction on various electrocatalysts.*

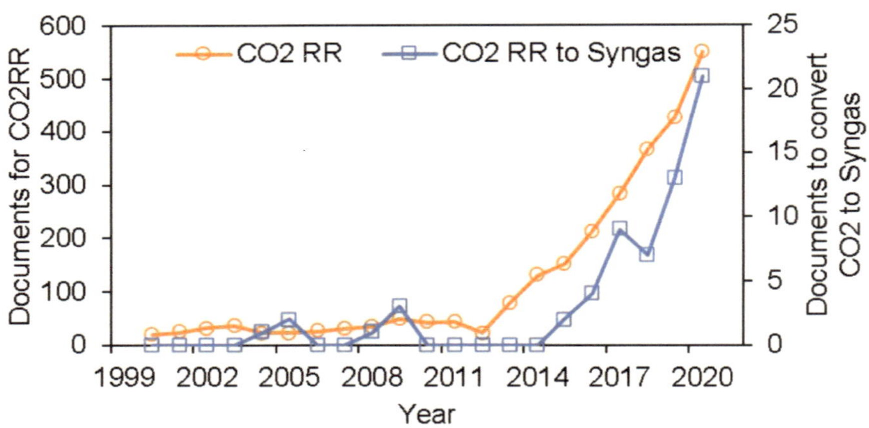


**Figure 3.**
*Total published documents for the electrochemical $CO_2$ reduction reaction and specifically convert $CO_2$ to SYNGAS in terms of over time [Scopus data based].*

there is not enough research in this field. Also, for the production of SYNGAS, the $Au_{0.76}$ –$Pd_{0.24}$ electrocatalyst has the highest Faraday efficiency (~90%) and CD (~10 $mA/cm^2$), which is a high-cost and unsuitable alloy electrode for large-scale use [42]. Other electrocatalysts for SYNGAS production have low FE and/or low CD, as can be seen in **Table 1**. The Ag/Au nanostructure catalysts for electrochemical $CO_2$RR to CO with a FE of further than 90% and a CD greater than 30 $mAcm^{-2}$ have been stated by researchers [40–43]. Zinc performs as an electrocatalyst for $CO_2$RR to CO, while it is a cost-effective, non-noble, and abundant choice to gold and silver [44].

There are also statements of nano-structured Zn catalysts including hexagonal, dendritic, and nanoscale [45–47]. Quan et al. have reported Zn nanoscale and Zn foil as a catalyst for the $CO_2$RR at the NaCl and $NaHCO_3$ electrolytes. They demonstrated that the nano-scale catalyst at NaCl cathodic solution has the greatest proficiency in terms of CD and FE about 6 mA.cm$^{-2}$ and > 90%, respectively, at a potential of −1.6 V by linear sweep voltammetry (LSV) method [45]. Rosen et al. have studied Zn balks and Zn dendrite catalysts for the electrochemical $CO_2$RR in 0.5 M $NaHCO_3$ cathodic solution. They stated Zn dendrite electrocatalyst has a CD of 4 mA.cm$^{-2}$ at the potential value of −1.14 V (*vs.* RHE) and FE of 80% [46]. By modifying the surface microstructure, morphology, or orientation of the Zn catalyst, the more FE and product selectivity can be attained for converting $CO_2$ to CO.

Nguyen et al. showed that microstructural or morphological changes in catalysts play a significant role in developing $CO_2$RR [48]. The surface of the Zn catalyst is simply oxidized although immersed in aqueous solutions or exposed to air. Thus, situations should be restricted to avoid zinc oxidation [48]. Nguyen et al. have also reported a porous nanostructure of the Zn catalysts which were prepared of zinc-oxide for the $CO_2$RR. By applying this porous metal, they obtained a faradaic efficiency of 78.5% for $CO_2$RR at a potential value of −0.95 V (*vs* RHE) in the $KHCO_3$ electrolyte [48]. Keerthiga and Chetty have reported a modified zinc-copper catalyst for the $CO_2$RR to hydrogen, $C_2H_6$, and $CH_4$ products. They coated zinc on the copper with different concentrations of electrolytes, and the outcomes were evaluated with pure Cu and Zn catalysts. They showed that zinc-copper with a high-level concentration of electrolyte had superior performance, also, the FE of $CH_4$ was the order Zn (7%) < Cu (23%) < Cu-Zn (52%). Moreover, the $H_2$ FE for Cu and Cu-Zn were 68% and 8%, respectively [49].

In this way, it has been selected inexpensive materials as electrocatalysts for commercial and industrial applications. Electrocatalysts must be appropriate that could have acceptable efficiency and cheap price for the reforming process. By referring to **Figure 2**, zinc and nickel are affordable materials for carbon monoxide and hydrogen production, respectively. Hence, to produce SYNGAS (CO + $H_2$) in this study, the Zn-Ni bimetallic material is chosen from these two groups of catalysts for CO and $H_2$ products. Other electrocatalysts are either inefficient or expensive. This work aims to investigate the $Zn_x$-$Ni_{1-x}$ coatings for the electrochemical $CO_2$ reduction reaction.

## 2. Experimental methods

### 2.1 Preparation and investigation of Zn-Ni Electrocatalyst for $CO_2$ reduction reaction

Zinc-nickel Alloys were coated on the low-carbon steel substrate by chronopotentiometry method at different electrochemical parameters. Then, Zn-Ni coatings were investigated in terms of microstructure, microcrack formation, and coating composition using SEM / EDX analysis and corrosion resistance by Autolab potentiostat (Model: PGSTAT128N) to obtain the coating with the best performance and quality. Besides, the coatings were analyzed using SEM/EDX analysis after $CO_2$ reduction reaction for microstructure and coke formation, as well as gas efficiency by gas chromatography analyzer. Nickel chloride hexahydrate ($NiCl_2.6H_2O$), ammonium chloride (NH4Cl), and zinc chloride ($ZnCl_2$) of raw materials were utilized for bath electrolyte preparation and ammonia solution (25%) for pH modification. All electrolytes were made using distilled water. The zinc and nickel alloy solutions were prepared in the laboratory to allow the study of the deposition at different bath solution temperatures. The pH of the solution was measured using a pH meter.

Ammonia solution (25%) was used to raise the pH of the electrolyte to the needed level of pH 5. The solution was stirred using a glass rod and the pH measuring was taken applying a pH meter, continuously. Chronopotentiometry electrodeposition was applied at different bath solution temperatures of 25°C, 40°C, 60°C, and 70°C. There were three types of electrodes, low carbon steel (working electrode), Ag/AgCl (reference electrode), and Pt mesh (counter electrode). The electrodeposition process was performed galvanostatically for each deposition temperature. The experimental setup for electrodeposition was performed as seen in **Figure 4**.

The deposited Zn-Ni coatings were analyzed on their compositional and microstructural properties applying SEM. The morphologies were observed and investigated for the electrodeposited zinc-nickel alloy samples at different temperatures of the bath solution. The material composition is determined by the SEM equipped with EDX which shows the composition information of the alloy coating. Linear polarization resistance (LPR) analysis was performed regarding the ASTM standard of G 96.– 90 (Reapproved 2001)e1.

### 2.2 Electrochemical $CO_2$ reduction reaction

For $CO_2$RR an H-shaped electrochemical cell was used which has 2-chambers (cathodic and anodic sections) that were connected with membrane Nafion 117 as seen in **Figure 5**. $CO_2$ gas was inserted into the cathodic section for the reduction process. In this method, electrocatalyst, reference electrodes (Ag/AgCl), and $CO_2$ saturated cathodic electrolyte were in the cathodic part, where $CO_2$RR happened, in the other part, the counter electrode (graphite) and anodic electrolyte (0.1 M $H_2SO_4$) were placed where the oxidation occurred. It was, therefore, predicted that SYNGAS ($H_2$ + CO) and coke would form in the cathodic portion, and $O_2$ would be produced in the anodic portion. Working, reference, and counter electrodes (WE, RE, and CE), were linked to the Autolab potentiostat device (Model: PGSTAT128N) to inspect potential and current records. A gas bag was attached to the exhaust to collect products for gas chromatography to characterize gases. Images of the catalyst morphologies were examined utilizing SEM. The catalysts were analyzed by EDX for coke formation and electrocatalyst surface compositions.

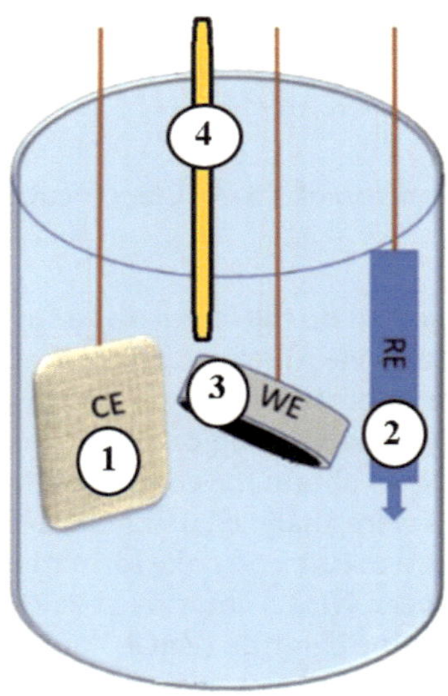


**Figure 4.**
*Electrodeposition setup for Zn-Ni coatings, (1) counter electrode, (2) reference electrode (3) working electrode, (4) thermometer.*

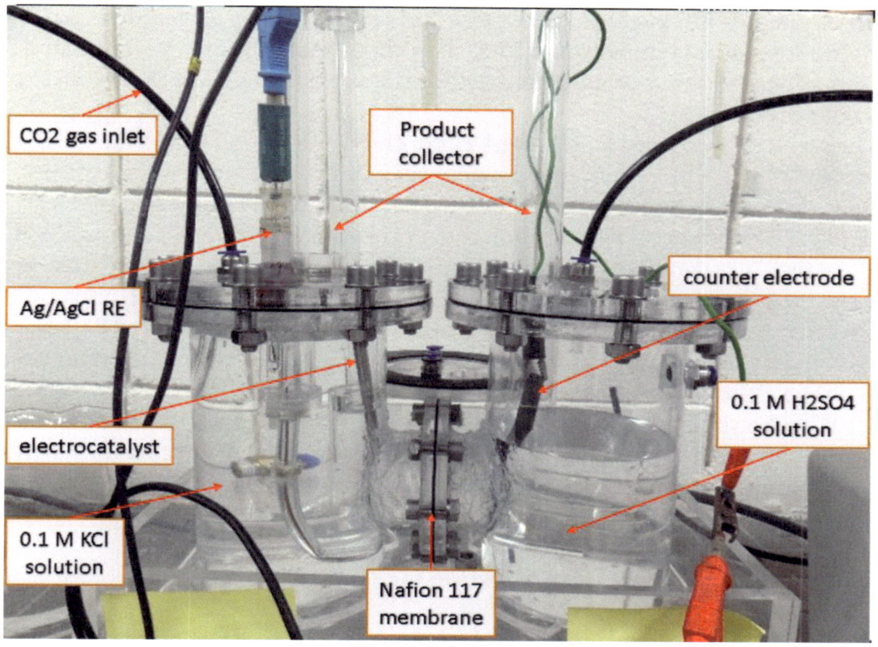


**Figure 5.**
*Cell setup for electrochemical $CO_2RR$ by Zn-Ni electrocatalyst, graphite CE, 0.1 M KCl (cathodic solution), $H_2SO_4$ (anodic solution) and Ag/AgCl RE.*

## 3. Results and discussions

### 3.1 Zn-Ni electrodeposition

The cathodic protection (CP) graph at various temperatures for the Zn-Ni deposition is displayed in **Figure 6**. The graph of the potential in terms of time for Zn-Ni coating depositions at 25°C, 40°C, 60°C, and 70°C were seen throughout

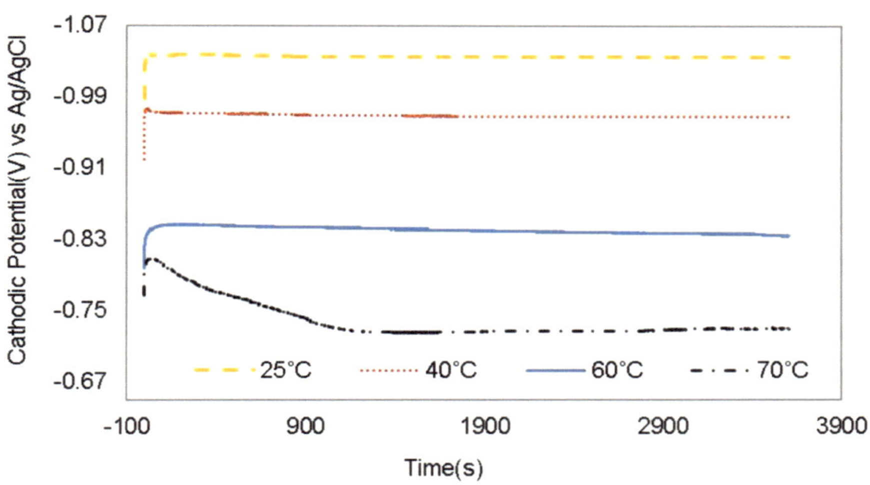


**Figure 6.**
*Deposition potential of Zn-Ni coatings in terms of time by chronopotentiometry method at various bath solution temperatures.*

the electrodeposition process. A decrease (more positive) in CP was detected over time with increasing temperature. The CP in chronopotentiometry was related to the ion's concentration becoming reduced at the substrate surface in response to the utilized current.

The standard potential $E_o$ (V) for Ni and Zn is −0.25V and 0.76 V (vs. SHE), respectively [90]. The CPs seen in the deposition were nearest to $E_o$ (V) of the reactants that were converted to its metal. Therefore, the outcomes on the decrease in CP towards a further positive amount over time for electrodepositions at high temperature (60°C and 70°C) demonstrated the CP's deposition was shifting nearer to $E_o$ (V) of Ni reduction, favoring the reaction of Ni-ion reduction to Ni-solid on the substrate. This opinion was more confirmed by EDX outcomes (**Figure 7**). The rise in Ni amount was assigned to a decrease in CP (more positive) over time through the electrodeposition reaction. This supposition is reported by Velichenko et al. [91], who stated the decrease in CP with rising Ni-ion concentration in the bath solution resulting in an improvement in Ni deposition. Qiao et al. [92] detected similar findings of reducing CP with increasing deposited Ni amount in the surface deposition with rising temperature.

### 3.2 Linear polarization resistance testing for Zn-Ni deposits

Zn and Ni amounts in the coatings have a considerable effect on the corrosion properties of Zn-Ni deposits. As revealed by Baldwin et al. [93] and Conde et al. [21], the lowest corrosion rate is obtained when the Ni amount is between 12 wt.% to 15 wt.% in the coating. Zn being a lower noble metal plays as an anode that sacrifices in relative to the substrate under a standard situation. Zn is extra favored compared with a nobler metal for instance Ni to be developed into coatings regarding its further sacrificial behavior. But adding more noble elements to Zn improves the corrosion resistance of Zn. By adding Ni to Zn, the rate of sacrificing of coating for the substrate is lower compared to bare Zn. Ni act to hinder or reduce the dissolution rate of Zn. But, when the Ni amount in the coating enhancements to more than 30%, the sacrificial performance decreases, and the coating turn nobler compared to the substrate (Steel). At this stage, the corrosion rate is entirely according to the coating characteristics. As seen in **Figure 8**, with rising temperatures, the corrosion rate increases. This shows that adding Ni to Zn no more enhances corrosion resistance. The coating turns nobler than the substrate and the existence of cracks that are detected causing a rise in the corrosion rate. As the bath solution temperature increases, hydrogen reduction occurs around the working electrode,

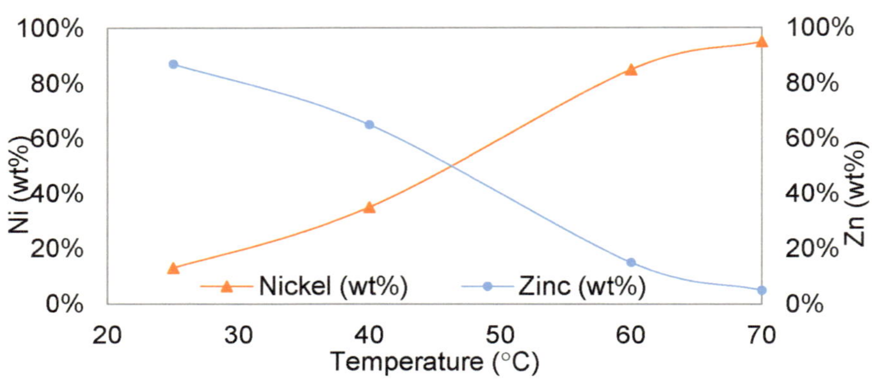


**Figure 7.**
*EDX analysis results in terms of Zn and Ni contents (wt%) vs. bath solution temperature (°C).*

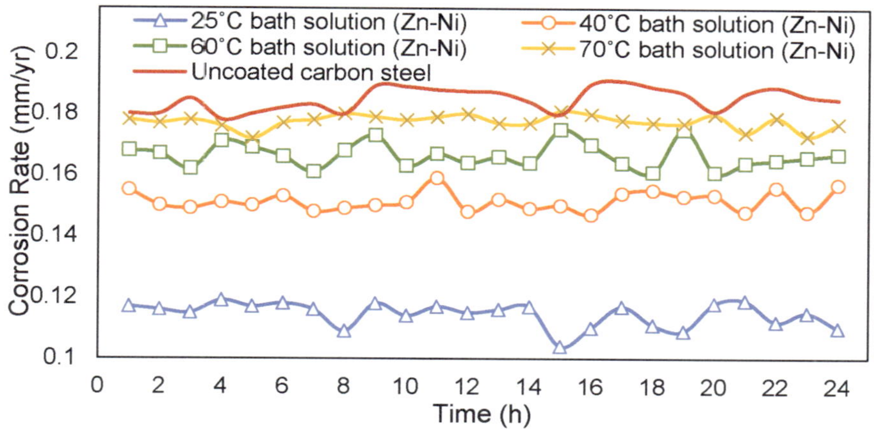


**Figure 8.**
*LPR corrosion rate measurements taken for Zn-Ni alloy coatings vs. uncoated carbon steel for hourly for 24 h.*

which creates bubbles form on the surface that prevents the deposition. On the other hand, hydrogen penetrates the coating and makes internal stress. The cracks and disruptions (as shown in **Figure 9**) in the coatings increase speed the corrosion rate of the substrate. These clarify the important variation in the corrosion rate for coatings deposited at 25°C and 40°C, 60°C, and 70°C.

The ratio of Zn and Ni for deposits formed at 25° C is in the optimal range of Ni amount from 12 to 15 wt.%. Therefore, the sacrificial performance of Zn is retained relative to the adding of the Ni, and this makes the steel substrate with decreasing corrosion rate as Zn acts as an anode. By adding 12–15 wt.% of Ni, the dissolution rate of Zn slows down, and the corrosion rate reduces. The cracks and defects in the

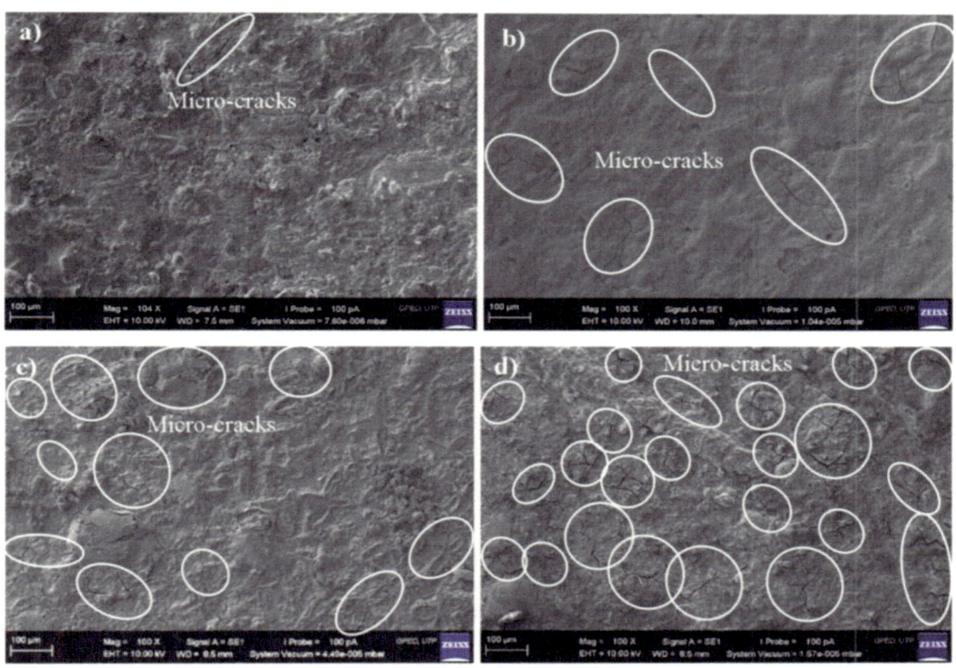


**Figure 9.**
*SEM images for electrodeposition of Zn-Ni alloy coatings at temperature of (a) 25°C, (b) 40°C, (c) 60°C, and (d) 70°C of bath solution.*

deposits do not substantially influence the corrosion properties of the metal layers, as further anodic Zn causes preferential corrosion.

### 3.3 SEM and EDX analysis for Zn-Ni deposits

As the bath electrolyte temperature raises, the ion mobility in the electrolyte rises. Hence, the coatings can be smoother. However, the SEM results displayed in **Figure 9** indicate that microcracks are detected in all deposited coatings at various temperatures. The micro-cracks intensity with rising the bath solution temperature is considered to be 25°C < 40°C < 60°C < 70°C. The microcracks formation can depend on the internal stress created and the evolution of hydrogen during the deposition. As the temperature increased, the evolution of hydrogen happened.

Enhancement of inner stress through deposition can be attributed to a lot of reasons. Alfantazi et al. [94] revealed the existence of microcracks in Zn-Ni coatings when the Ni amount in the coating increased. Qiao et al. [92] and Rehim et al. [95] reported the micro-cracks in the Zn-Ni coatings deposited in the acidic bath solution were attributed to $H_2$ embrittlement via the $H_2$ evolution. A rise in the hydrogen release was observed with the outputs of a rise in the hydrogen CD with the temperature rises. This $H_2$ reduction reaction causes $H_2$ atoms to penetrate the coated layer, straining the crystal lattice, and causing high-stress internal cracks.

### 3.4 Investigation of Zn-Ni bimetallic electrocatalysts for $CO_2$RR

To realize the impacts of the catalysts for the $CO_2$RR, the composition, morphology, and structure of the catalysts were investigated. **Figure 10** and **Figure 11** display EDX results and SEM images of the Zn-Ni with various compositions after the 48 h for the $CO_2$RR by cyclic voltammetry with scan rate 0.05 V. $s^{-1}$, graphite counter electrode, 0.1 M KCl cathodic, and 0.1 M $H_2SO_4$ anodic solutions. According to EDX analysis, as shown in **Figure 10**, carbon with ~28–30 wt.% was deposited on the $Zn_{85\%}$-$Ni_{15\%}$ electrocatalyst after 48 h of testing. The microstructure of $Zn_{0.85}$ - $Ni_{0.15}$ is a block-like morphology in which carbon is almost uniformly distributed in the substrate due to $CO_2$ reduction. As can be seen in **Figure 11(a)**, some electrocatalytic regions are carbon-covered, preventing $CO_2$ reduction over time. Therefore, for further consideration of this electrocatalyst, gas chromatography of produced gases (the produced gases were collected with the gas bag) has been investigated.

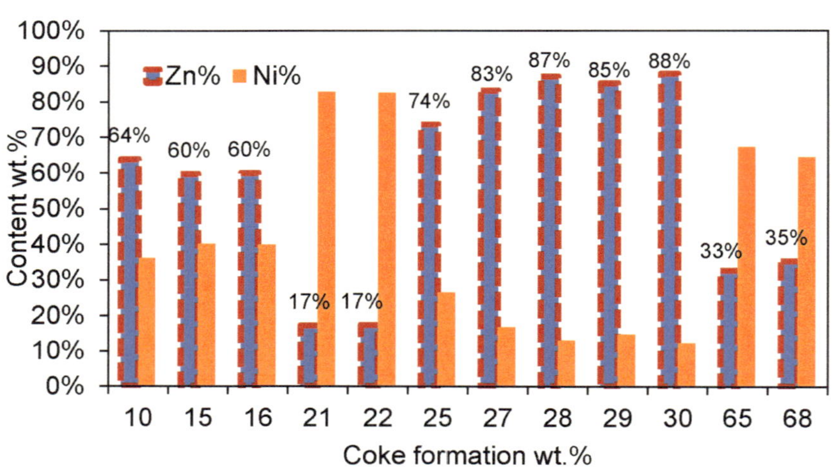


**Figure 10.**
*EDX results of C content (wt.%) in terms of Zn-Ni compositions after 48 h of electrochemical $CO_2$RR.*

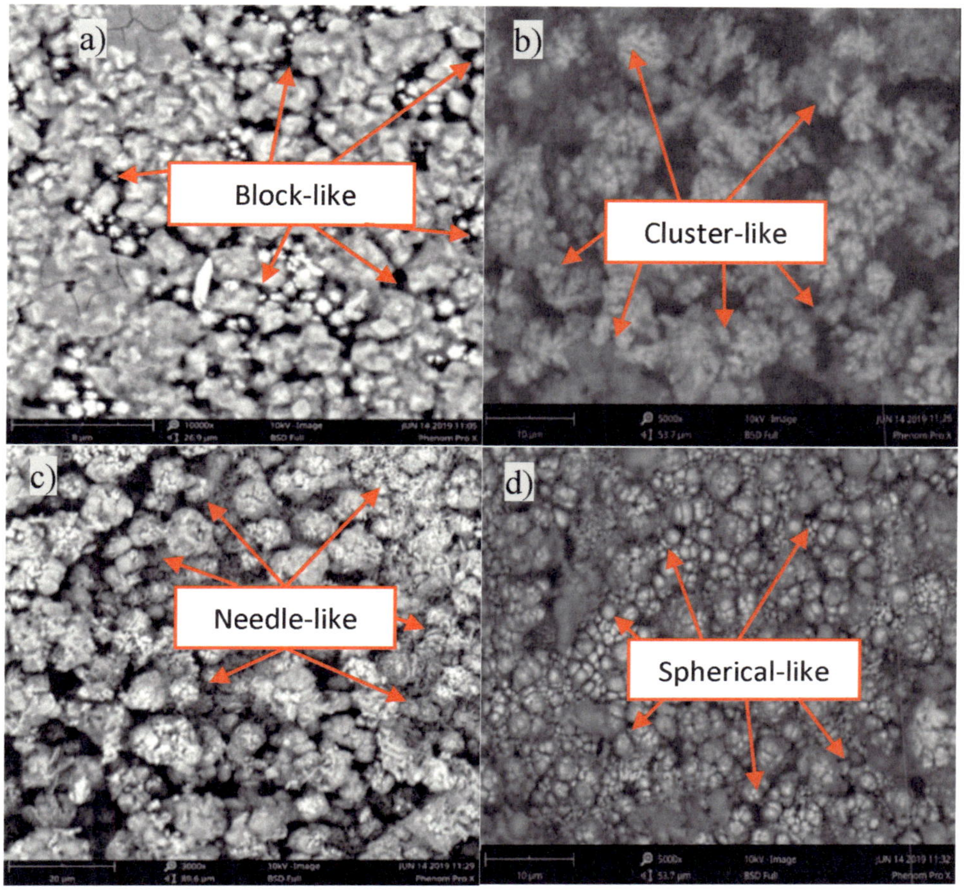


**Figure 11.**
*SEM images of Zn-Ni electrocatalysts after 48 h of $CO_2$RR on (a) $Zn_{85\%}$- $Ni_{15\%}$, (b) $Zn_{65\%}$-$Ni_{35\%}$, (c) $Zn_{35\%}$-$Ni_{65\%}$, and (d) $Zn_{20\%}$-$Ni_{80\%}$ electrocatalysts.*

According to EDX analysis, as shown in **Figure 10**, carbon with ~10 wt.% was deposited on the $Zn_{65\%}$-$Ni_{35\%}$ electrocatalysts after 48 h of testing. As shown in **Figure 11(b)**, the microstructure of the $Zn_{65\%}$-$Ni_{35\%}$ electrocatalyst is a cluster-like morphology where coke formation is minimized by the reaction of $CO_2$ with this microstructure after 48 h. With comparing $Zn_x$-$Ni_{1-x}$ electrocatalysts, with decreasing Zn amount in $Zn_x$-$Ni_{1-x}$ coatings from 85 wt.% to 65 wt.% of Zn, coke formation upon Zn-Ni electrocatalysts decreases. Furthermore, the electrocatalyst microstructures have changed from block-like to cluster-like with decreasing Zn content from 85 wt.% to 65 wt.%. Therefore, the activity and efficiency of electrocatalysts increase with decreasing Zn content from 85 wt.% to 65 wt.% in Zn-Ni electrocatalysts. By further reducing the amount of Zn until ~33 wt.%, coke formation upon Zn-Ni electrocatalyst increases.

Furthermore, as seen in **Figure 11c**, the microstructure of the $Zn_{35\%}$-$Ni_{65\%}$ electrocatalyst is semi-spherical, where carbon was deposited between the semi-spherical grains with needle-like microstructure. This high coke formation is due to changes in the microstructure and electrocatalytic activity due to the interaction between Zn and Ni with the ions present in the solution. By further reducing the amount of Zn up to 20 wt.% in the Zn-Ni coating, the coke formation (after 48 h of $CO_2$RR) on the electrocatalyst decreased. The microstructure of 20%Zn-80%Ni is a glossy spherical morphology where carbon is grown with a dark semi-spherical morphology about 22 wt.%.

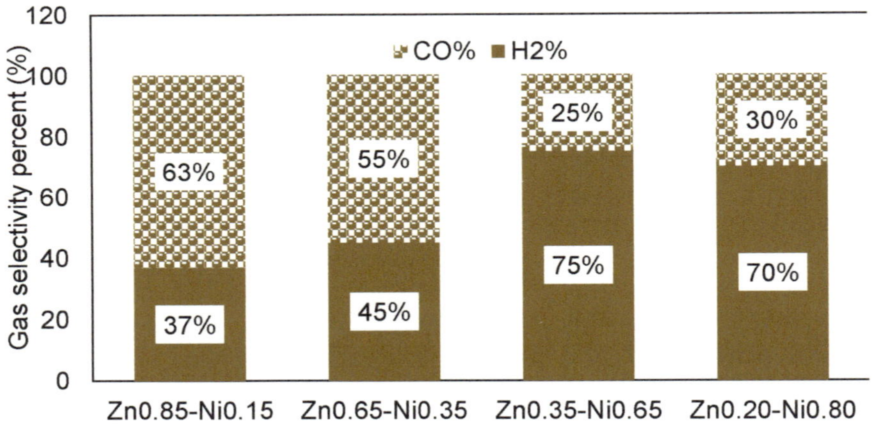


**Figure 12.**
*Gas chromatography results for $CO_2RR$ by various electrocatalysts in terms of CO and $H_2$ gas selectivity by cyclic voltammetry method.*

| Coating composition | | | | | | Coating performance | | | |
|---|---|---|---|---|---|---|---|---|---|
| | Zn content (%) | Ni content (%) | CR (mm/yr) | Morphology | | CO (%) | $H_2$ (%) | C (%) | Efficiency after 48 h (%) |
| | | | | Coating | carbon | | | | |
| $Zn_x$-$Ni_{1-x}$ | 85 | 15 | 0.11 | Block-white | Block-black | 63 | 37 | 30 | 53.33 |
| | **65** | **35** | **0.15** | **Cluster-like** | **Low amount** | **55** | **45** | **10** | **66.00** |
| | 35 | 65 | 0.17 | Semi-spherical | Needle-like | 25 | 75 | 65 | 31.88 |
| | 20 | 80 | 0.25 | White Spherical | black Spherical | 30 | 70 | 22 | 57.27 |

**Table 2.**
*Zn-Ni Electrocatalysts performance with different compositions and electrodeposition parameters for $CO_2RR$.*

Due to the results of gas chromatography, as shown in **Figure 12**, the $Zn_{85\%}$-$Ni_{15\%}$, $Zn_{65\%}$-$Ni_{35\%}$, $Zn_{35\%}$-$Ni_{65\%}$, and $Zn_{20\%}$-$Ni_{80\%}$ electrocatalysts have 63%, 55%, 25%, and 30% selectivity for CO and 37%, 45%, 75%, and 70% selectivity for $H_2$ products, respectively. Also, according to **Table 2**, the total efficiency for $CO_2RR$ after 48 h of testing is 53%, 66%, 31%, and 57%, respectively. The $Zn_{65\%}$-$Ni_{35\%}$ electrocatalyst is appropriate in terms of morphology, stability, coke formation, product selectivity, and intensity of the electrochemical $CO_2RR$. The coke formation on the catalysts can influence the activity spots of the catalyst and have a negative impact on the efficiency and life cycle of the catalyst. Consequently, the chemical compositions, microstructure, and morphology of catalysts have a crucial role for the $CO_2RR$ to produce gases with satisfactory ratio, desired product, least-coke formation, and suitable efficiency, activity, and stability.

## 4. Conclusions

The lower corrosion rate of coatings deposited at 25°C is mainly related to the role of nickel in zinc-nickel alloy and a higher corrosion rate at higher temperatures of 40°C, 60°C, and 70°C are related to the lower barrier properties such as uniformity, compactness and cracks in the alloy. Zinc-nickel alloy coatings with the highest corrosion resistance, within required composition of 12–15%, dense and compact morphology, better uniformity with less crack is achieved with coatings deposited at 25°C. $CO_2$RR on $Zn_x$-$Ni_{1-x}$ electrocatalysts in 0.1 M KCl as the cathodic solution and 0.1 M $H_2SO_4$ as the anodic solution using cyclic voltammetry method demonstrated that the $Zn_{65\%}$-$Ni_{35\%}$ electrode had the best performance for the $CO_2$RR with regarding the minimum coke formation (<10%) and optimal faradaic efficiencies of CO and $H_2$ ($FE_{CO}$ = 55% and $FE_{H2}$ = 45%). The coke formation on the catalysts can influence the activity spots of the catalyst and have a negative impact on the efficiency and life cycle of the catalyst. Consequently, the chemical compositions, microstructure, and morphology of catalysts have a crucial role for the $CO_2$RR to produce gases with satisfactory ratio, desired product, least-coke formation, and suitable efficiency, activity, and stability. Therefore, the $Zn_{65\%}$-$Ni_{35\%}$ electrocatalyst with cluster microstructure had the best performance for $CO_2$RR among other electrocatalysts in this study.

## Author details

Mohammadali Beheshti*, Saeid Kakooei, Mokhtar Che Ismail and Shohreh Shahrestani
Universiti Teknologi PETRONAS, Seri Iskandar, Malaysia

*Address all correspondence to: mohammadali_g03438@utp.edu.my

## References

[1] Yua F, Weia P, Yanga Y, Chena Y, Guob L, Peng Z. Material design at nano and atomicscale for electrocatalytic $CO_2$ reduction. Nano material. Science. 2019;1:60-69.

[2] Kortlever R, Balemans C, Kwon Y, Koper MTM. Electrochemical $CO_2$ reduction to formic acid on a Pd-based formicacid oxidation catalyst. Catalyst Today. 2015;244:58-62.

[3] Liu Y, Tian D, Biswas AN, Xie Z, Hwang S, Lee JH, Meng H, Chen JG. Transition Metal Nitrides as Promising Catalyst Supports for Tuning $CO/H_2$ Syngas Production from Electrochemical $CO_2$ Reduction. Angewandte Chemie. 2020;131. DOI: https://doi.org/10.1002/ange.202003625

[4] Derrick J, Loipersberger M, Iovan D, Smith PT, Chakarawet K, Long JR, Head-Gordon M, Chang C. Metal–Ligand Exchange Coupling Promotes Iron-Catalyzed Electrochemical $CO_2$ Reduction at Low Overpotentials. ChemRxiv. 2020. DOI: https://doi.org/10.26434/chemrxiv.11923176.v1

[5] Beheshti M, Kakooei S, Ismail MC, Shahrestani S. Investigation of $CO_2$ electrochemical reduction to Syngas on Zn/Ni-based electrocatalysts using the cyclic voltammetry method. Electrochimica Acta. 2020;341:135976. DOI: https://doi.org/10.1016/j.electacta.2020.135976

[6] Ma Z, Tsounis C, Kumar PV, Han Z, Wong RJ, Toe CY, Zhou S, Bedford NM, Thomsen L, Ng YH, Amal R. Enhanced Electrochemical $CO_2$ Reduction of Cu@Cux O Nanoparticles Decorated on 3D Vertical Graphene with Intrinsic $sp^3$-type Defect. Advanced Functional Materials. 2020: 1910118. DOI: https://doi.org/10.1002/adfm.201910118.

[7] Choi W, Won DH, Hwang YJ. Catalyst design strategies for stable electrochemical $CO_2$ reduction reaction. Journal of Materials Chemistry A. 2020;120. https://doi.org/10.1039/D0TA02633F

[8] Perry SC, Leung P, Wang L, Ponce de León C. Developments on carbon dioxide reduction: Their promise, achievements, and challenges. Current Opinion in Electrochemistry. 2020;20:88-98,2020. DOI:https://doi.org/10.1016/j.coelec.2020.04.014

[9] Zhao Y, Chang X, Malkani AS, Yang X, Thompson L, Jiao F, Xu B. Speciation of Cu Surfaces During the Electrochemical CO Reduction Reaction. Journal of the American Chemical Society. 2020;142(21):9735-9743. DOI:https://doi.org/10.1021/jacs.0c02354

[10] Zhang L, Merino-Garcia I, Albo J, Sánchez-Sánchez CM. Electrochemical $CO_2$ reduction reaction on cost-effective oxide-derived copper and transition metal–nitrogen–carbon catalysts. Current Opinion in Electrochemistry. 2020;23:65-73.

[11] Zhang T, Han X, Yang H, Han A, Hu E, Li Y, Yang X, Wang L, Liu J, Liu B. Atomically Dispersed Nickel(I) on an Alloy-Encapsulated Nitrogen-Doped Carbon Nanotube Array for High-Performance Electrochemical $CO_2$ Reduction Reaction. Chemistry Europe. 2020. DOI: https://doi.org/10.1002/ange.202002984

[12] Garg S., Li M., Rufford TE, Ge L, Rudolph V, Knibbe R, Konarova M, Wang GGX. Catalyst–Electrolyte Interactions in Aqueous Reline Solutions for Highly Selective Electrochemical $CO_2$ Reduction. 2020. DOI: https://doi.org/10.1002/cssc.201902433

[13] Popović S, Smiljanić M, Jovanovič P, Vavra J, Buonsanti R, Hodnik N. Stability and Degradation Mechanisms of Copper-Based Catalysts for

Electrochemical $CO_2$ Reduction. chemical Society. 2020. DOI:https://doi.org/10.1002/anie.202000617

[14] Gunathunge CM, Li J, Li X, Hong JJ, Waegele MM. Revealing the Predominant Surface Facets of Rough Cu Electrodes under Electrochemical Conditions. ACS Catalysis. 2020;10(12):6908-6923. DOI:https://doi.org/10.1021/acscatal.9b05532

[15] Yu H, Jiang J., Accelerating catalysts design by machine learning. Science Bulletin. 2020. DOI: https://doi.org/10.1016/j.scib.2020.06.004

[16] Zhi X, Jiao Y, Zheng Y, Vasileff A, Qiao SZ. Selectivity roadmap for electrochemical $CO_2$ reduction on copper-based alloy catalysts. Nano Energy. 2020;7(1):104601.

[17] Rutkowska IA, Wadas A, Szaniawska E, Chmielnicka A, Zlotorowicz A, Kulesza P J. Elucidation of activity of copper and copper oxide nanomaterials for electrocatalytic and photoelectrochemical reduction of carbon dioxide. Current Opinion in Electrochem. 2020. DOI:https://doi.org/10.1016/j.coelec.2020.05.014

[18] Zhou Y, Zhou R, Zhu X, Han N, Song B, Liu T, Hu G, Li Y, Lu J, Li Y. Mesoporous PdAg Nanospheres for Stable Electrochemical $CO_2$ Reduction to Formate. Advanced Materials. 2020. DOI:https://doi.org/10.1002/adma.202000992

[19] Barcelo G, Sarret M, Müller C, Pregonas J. Corrosion resistance and mechanical properties of zinc electrocoatings. Electrochimical Acta. 1998;43:13-20.

[20] Bowden C, Matthews A. A study of the corrosion properties of PVD Zn-Ni coatings. Surface and Coatings Technology. 1995;76:508-515.

[21] Conde A, Arenas M, De Damborenea J. Electrodeposition of Zn–Ni coatings as Cd replacement for corrosion protection of high strength steel. Corrosion Science. 2011;53: 1489-1497.

[22] Beheshti M, Ismail MC, Kakooei S, Shahrestani S. Influence of temperature and potential range on Zn-Ni deposition properties formed by cyclic voltammetry electrodeposition in chloride bath solution. Corrosion Reviews.2020;38(2):127-136. DOI: https://doi.org/10.1515/corrrev-2019-0086

[23] Lin Zf, Li Xb, Xu Lk. Electro-deposition and corrosion behavior of zinc–nickel films obtained from acid solutions: effects of TEOS as additive. International Journal of Electrochemistry science. 2012;7:12507-12517.

[24] Beheshti M, Ismail MC, Kakooei S, Shahrestani S, Mohan G, Zabihiazadboni M. Influence of deposition temperature on the corrosion resistance of electrodeposited zinc-nickel alloy coatings. Materialwissenschaft und Werkstofftechnik. 2018;49(4):472-482. DOI: https://doi.org/10.1002/mawe.201700284

[25] Byk T, Gaevskaya T, Tsybulskaya L. Effect of electrodeposition conditions on the composition, microstructure, and corrosion resistance of Zn–Ni alloy coatings. Surface and Coatings Technology. 2008;202:5817-5823.

[26] D'Alessandro DM, Smit B, Long JR. Carbon dioxide capture: prospects for new materials. Angewandte Chemie International Edition, 2010;49:6058-6082.

[27] Hori Y. Electrochemical $CO_2$ reduction on metal electrodes. In: Modern Aspects of Electrochemistry. Heidelberg: Springer; 2008. 89-189 pp.

[28] Qiao JL, Liu YY, Hong F, Zhang JJ. A review of catalysts for the electroreduction of carbon dioxide to produce low-carbon fuels. Chemical Society Reviews. 2014;43:631-675.

[29] Jhong HR, Ma S, Kenis PJ. Electrochemical conversion of $CO_2$ to useful chemicals: current status, remaining challenges, and future opportunities. Current Opinion in Chemical Engineering. 2013;2:191-199.

[30] Lee J, Kwon Y, Machunda RL, Lee HJ. Electrocatalytic recycling of $CO_2$ and small organic molecules. Asian Journal of Chemistry, 2009;4:1516-1523.

[31] Durand WJ, Peterson AA, Studt F, Abild-Pedersen F, Nørskov JK. Structure effects on the energetics of the electrochemical reduction of $CO_2$ by copper surfaces. Surface Science. 2011;605:1354-1359.

[32] Hansen HA, Montoya JH, Zhang YJ, Shi C, Peterson AA, Nørskov JK. Electroreduction of methanediol on copper. Catalysis Letters. 2013;143:631-635.

[33] Peterson AA, Abild-Pedersen F, Studt F, Rossmeisl J, Nørskov JK. How copper catalyzes the electroreduction of carbon dioxide into hydrocarbon fuels. Energy and Environmental Sciences. 2010;3:1311-1315.

[34] Peterson AA, Nørskov JK. Activity descriptors for $CO_2$ electroreduction to methane on transition-metal catalysts. The Journal of Physical Chemistry Letters. 2012;3:251-258.

[35] Cui C, Wang H, Zhu X, Han J, Ge Q. A DFT study of $CO_2$ electrochemical reduction on Pb(211) and Sn(112). Science China Press and Springer-Verlag Berlin Heidelberg. 2015;58(4):607-613.

[36] Lu Q, Jiao F. Electrochemical $CO_2$ reduction: Electrocatalyst, reaction mechanism, and process engineering. Nano Energy. 2016;4-9:2211-2855.

[37] Hirunsit P. Electroreduction of carbon dioxide to methane on copper, copper-silver, and copper-gold catalysts: a DFT study Journal of Physical Chemistry C. 2013;117: 8262-8268.

[38] Nie X, Esopi MR, Janik MJ, Asthagiri A. Selectivity of $CO_2$ reduction on copper electrodes: the role of the kinetics of elementary steps. Angewandte Chemie International Edition. 2013;52:2459-2462.

[39] Nie X, Luo W, Janik MJ, Asthagiri A. Reaction mechanisms of $CO_2$ electrochemical reduction on Cu (111) determined with density functional theory. Journal of Catalyst. 2014;312:108-122.

[40] Chen Y, Li CW, Kanan MW. Aqueous $CO_2$ Reduction at Very Low Overpotential on Oxide-Derived Au Nanoparticles. Journal of the American Chemical Society. 2012;134(49):19969-19972.

[41] Rosen J, Hutchings GS, Lu Q, Rivera S, Zhou Y, Vlachos DG, Jiao F. Mechanistic Insights into the Electrochemical Reduction of $CO_2$ to CO on Nanostructured Ag Surfaces. ACS Catalysis. 2015;5(7):4293-4299.

[42] Nursanto EB, Jeon HS, Kim C, Jee MS, Koh JH, Hwang YJ, Min BK. Gold catalyst reactivity for $CO_2$ electroreduction: From nano particle to layer. Catalysis Today. 2016;260:107-111.

[43] Jee MS, Jeon HS, Kim C, Lee H, Koh JH, Cho J, Min BK, Hwang YJ. Enhancement in carbon dioxide activity and stability on nanostructured silver electrode and the role of oxygen. Applied Catalysis B. 2016;180:372-378.

[44] Hori Y, Vayenas CG, White RE, Gamboa-Aldeco ME. Electrochemical $CO_2$ Reduction on Metal Electrodes. MOD. ASPECT. ELECTROC. 2008;42: 89-104.

[45] Quan F, Zhong D, Song H, Jia F, Zhang L. A highly efficient zinc catalyst for selective electroreduction of carbon dioxide in aqueous NaCl solution. Journal of Materials Chemistry A. 2015;3(32):16409-16413.

[46] Rosen J, Hutchings GS, Lu Q, Forest RV, Moore A, Jiao F. Electrodeposited Zn Dendrites with Enhanced CO Selectivity for Electrocatalytic $CO_2$ Reduction. ACS Catalysis. 2015;5(8):4586-4591.

[47] Won DH, Shin H, Koh J, Chung J, Lee HS, Kim H, Woo SI. Highly Efficient, Selective, and Stable $CO_2$ Electroreduction on a Hexagonal Zn Catalyst. Angewandte Chemie International Edition. 2016;55(32):9297-9300.

[48] Nguyen DLT, Jee MS, Won DH, Jung H, Oh HS, Min BK, Hwang YJ. Selective $CO_2$ Reduction on Zinc Electrocatalyst: The Effect of Zinc Oxidation State Induced by Pretreatment Environment. ACS Sustainable Chemistry & Engineering. 2017;5:11377-11386.

[49] Keerthiga G, Chetty R. Electrochemical Reduction of Carbon Dioxide on Zinc-Modified Copper Electrodes. Journal of the Electrochemical Society. 2017;164(4):H164-H169.

[50] Hori Y, Murata A. Electrochemical evidence of intermediate formation of adsorbed CO in cathodic reduction of $CO_2$ at a nickel electrode. Electrochim. Acta. 1990;35:1777-1780.

[51] Murata A, Hori Y. Electrochemical reduction of $CO_2$ to CO at Ni electrodesmodified with CD. Chemical Letter. 1991:181-184.

[52] Hori Y, Kikuchi K, Suzuki S. Production of CO and $CH_4$ in electrochemicalreduction of $CO_2$ at metal–electrodes in aqueous hydrogencarbonatesolution. Chemical Letter. 1985:1695-1698.

[53] Taguchi S, Aramata A. Surface-structure sensitive reduced $CO_2$ formationon Pt single crystal electrodes in sulfuric acid solution. Electrochimical Acta, 1994;39:2533-2537.

[54] Kumar B, Brian JP, Atla V, Kumari S, Bertram KA, White RT, Spurgeon JM. New trends in the development of heterogeneous catalysts for electrochemical $CO_2$ reduction. Catalysis Today. 2016;270:19-30.

[55] Jhong HRM, Ma S, Kenis PJ. Electrochemical conversion of $CO_2$ to useful chemicals: current status, remaining challenges, and future opportunities. Current Opinion in Chemical Engineering. 2013;2(2): 191-199.

[56] Jhong HR, Brushett FR, Yin L, Stevenson DM, Kenis PJA. Combining Structural and Electrochemical Analysis of Electrodes Using Micro-Computed Tomography and a Microfluidic Fuel Cell. Journal of Electrochemical Society. 2012;159(3):292-298.

[57] Cindrella L, Kannan AM, Lin JF, Saminathan K, Ho Y, Lin CW, Wertz J. Gas diffusion layer for proton exchange membrane fuel cells—A review. PowerSources. 2009;194:146-160.

[58] Furuya N, Yamazaki T, Shibata M. High performance Ru/Pd catalysts for $CO_2$ reduction at gas-diffusion electrodes. Journal of Electroanalytical Chemistry. 1997;431:39-41.

[59] Rosen BA, Zhu W, Kaul G, Khojin AS, Masel RI. Water Enhancement of $CO_2$ Conversion on Silver in 1-Ethyl-3-Methylimidazolium Tetrafluoroborate. Journal of Electrochemical Society. 2013;160:138-141.

[60] Rosen BA, Khojin AS, Thorson MR, Zhu W, Whipple DT, Kenis PJA, Masel RI. Ionic Liquid–Mediated Selective Conversion of $CO_2$ to CO at Low Overpotentials. Science. 2011;334:643-644.

[61] Ramos JM, Pupillo RC, Keane TP, DiMeglio JL, Rosenthal J. Efficient Conversion of $CO_2$ to CO Using Tin and Other Inexpensive and Easily Prepared Post-Transition Metal Catalysts. Journal of the American Chemical Society. 2015;137:5021-5027.

[62] Asadi M, Kumar B, Behranginia A, Rosen BA, Baskin A, Repnin N, Pisasale D, Phillips P, Zhu W, Haasch R, Klie RF, Kra'l P, Abiade J, Khojin AS. Robust carbon dioxide reduction on molybdenum disulphide edges. Nature Communications. 2014;5:1-8.

[63] Oh Y, Vrubel H, Guidoux S, Hu X. Efficient Reduction of $CO_2$ to CO with High Current Density Using in Situ or ex Situ Prepared Bi-Based Materials. Chemical Communications. 2014;50:3878-3881.

[64] Oh Y, Hu X. Ionic liquids enhance the electrochemical $CO_2$ reduction catalyzed by $MoO_2$. Chemical Communications. 2015;51:13698-13701.

[65] DiMeglio JL, Rosenthal J. Selective Conversion of $CO_2$ to CO with High Efficiency Using an Inexpensive Bismuth-Based Electrocatalyst. Journal of the American Chemical Society. 2013;135:8798-8801.

[66] Ganesh I, Kumar PP, Annapoorna I, SumlinerJM,RamakrishnaM,HebalkarNY, Padmanabham G, Sundararajan G. Preparation and characterization of Cu-doped $TiO_2$ materials for electrochemical, photoelectrochemical, and photocatalytic applications. Applied Surface Science. 2014;293:229-247.

[67] Genovese C, Ampelli C, Perathoner S, Centi G. Electrocatalytic conversion of $CO_2$ on carbon nanotube-based electrodes for producing solar fuels. Journal of catalyst. 2013;308: 237-249.

[68] Bernstein NJ, Akhade SA, Janik MJ. Density functional theory study of carbon dioxide electrochemical reduction on the Fe(100) surface. Physical Chemistry Chemical Physics. 2014;16:13708-13717.

[69] Friebel D, Mbuga F, Rajasekaran S, Miller DJ, Ogasawara H, Alonso -Mori R, Sokaras D, Nordlund D, Weng TC, Nilsson A. Structure, redox chemistry, and interfacial alloy formation in monolayer and multilayer Cu/Au(111) model catalysts for $CO_2$ electroreduction. The Journal of Physical Chemistry C. 2014;118:7954-7961.

[70] Terunuma Y, Saitoh A, Momose Y. Relationship between hydrocarbon production in the electrochemical reduction of $CO_2$ and the characteristics of the Cu electrode. Journal of Electroanalytical Chemistry. 1997;434: 69-75.

[71] Goncalves M., Gomes A, Condeco J, Fernandes R, Pardal T, Sequeira C, Branco J. Selective electrochemical conversion of $CO_2$ to $C_2$ hydrocarbons. Energy Convers Manage. 2010;51: 30-32.

[72] Schouten KJP, Kwon Y, van der Ham CJM., Qin Z, Koper MTM. A new mechanism for the selectivity to C-1 and C-2 species in the electrochemical reduction of carbon dioxide on copper electrodes. Chemical Science. 2011;2:1902-1909.

[73] Hori Y, Takahashi I, Koga O, Hoshi N. Selective formation of $C_2$ compounds from electrochemical reduction of $CO_2$ at a series of copper single crystal electrodes. Journal of Physical Chemistry B. 2002;106:15-17.

[74] Hori Y, Konishi H, Futamura T, Murata A, Koga O, Sakurai H, Oguma K. Deactivation of copper electrode" in electrochemical reduction of $CO_2$, Electrochimical Acta. 2005;50: 5354-5369.

[75] Fukatsu A, Kondo M, Okabe Y, Masaoka S. porphyrin-catalyzed $CO_2$ reduction under photoirradiation. Journal

of Photochemistry and Photobiology A: Chemistry. 2015;313:143-148.

[76] Yang W, Dastafkan K, Jia C, Zhao C. Design of Electrocatalysts and Electrochemical Cells for Carbon Dioxide Reduction Reactions. Advanced Material Technology. 2018;1700377. DOI: 10.1002/admt.201700377

[77] Torelli DA, Francis SA, Crompton JC, Javier A, Thompson JR, Brunschwig BS, Soriaga MP, Lewis NS. Nickel-Gallium-Catalyzed Electrochemical Reduction of $CO_2$ to Highly Reduced Products at Low Overpotentials. ACS Catalysis. 2016; 6( 3):2100-2104.

[78] Ampelli C, Genovese C, Perathoner S, Centi G, Errahali M, Gatti G, Marchese L. An Electrochemical Reactor for the $CO_2$ Reduction in Gas Phase by Using Conductive Polymer Based Electrocatalysts. Chemical Engineering Transactions. 2014;41:13-18. DOI: 10.3303/cet 1441003,

[79] Mignard D, Barik RC, Bharadwaj AS, Pritchard CL, Ragnoli M, Cecconi F, Miller H, Yellowlees LJ. Revisiting strontium-doped lanthanum cuprate perovskite for the electrochemical reduction of $CO_2$. Journal of $CO_2$ Utilization. 2014;5:53-59.

[80] Ullah N, Ali I, Jansen M, Omanovic S. Electrochemical Reduction of $CO_2$ in an Aqueous Electrolyte Employing an Iridium/Ruthenium-Oxide Electrode. Canadian Journal of Chemical Engineering. 2015;93:55-62. DOI 10.1002/cjce.22110,

[81] Guo X, Zhang Y, Deng C, Li X, Xue Y, Yan YM, Sun K. Composition dependent activity of Cu–Pt nanocrystals for electrochemical reduction of $CO_2$. Chem. Commun. 2015;51:1345-1352.

[82] Maark TA, Nanda BRK. CO and $CO_2$ Electrochemical Reduction to Methane on Cu, Ni, and $Cu_3Ni$ (211) Surfaces. J. Phys. Chem. C. 2016;120:8781-8789.

[83] Pérez-Rodrígueza S, Rilloa N, Lázaroa MJ, Pastor E. Pd catalysts supported onto nanostructured carbon materials for $CO_2$ valorization by electrochemical reduction. Applied Catalysis B: Environmental. 2015;163:83-95.

[84] Bellini M, Folliero MG, Evangelisti C, He Q, Hu Yf, Pagliaro MV, Oberhauser W, Marchionni A, Filippi J, Miller HA, Vizza F. A Gold-Palladium Nanoparticle Alloy Catalyst for CO Production from $CO_2$ Electroreduction. Energy Technology. 2018.https://doi.org/10.1002/ente.201800859

[85] Grote JP, Zeradjanin AR, Cherevko S, Savan A, Breitbach B, Ludwig A, Mayrhofer KJJ. Screening of material libraries for electrochemical $CO_2$ reduction catalysts – Improving selectivity of Cu by mixing with Co. Journal of Catalysis.2016;343:248-256.

[86] Ma M, Hansen HA, Valenti M, Wang Z, Cao A, Dong M, Smith WA. Electrochemical reduction of $CO_2$ on compositionally variant Au-Pt bimetallic thin films. Nano Energy. 2017; 42:51-57.

[87] Choi J, Kim MJ, Ahn SH, Choi I, Jang JH, Ham YS, Kim JJ, Kim SK. Electrochemical $CO_2$ reduction to CO on dendritic Ag–Cu electrocatalysts prepared by electrodeposition. Chemical Engineering Journal, 2016; 299:37-44.

[88] Zhu Q, Ma J, Kang X, Sun X, Hu J, Yang G, Han B. Electrochemical reduction of $CO_2$ to CO using graphene oxide/carbon nanotube electrode in ionic liquid/acetonitrile system., Chemistry. 2016;59-5: 551-556. doi: 10.1007/s11426-016-5584-1

[89] Aeshala LM, Uppaluri RG, Verma A. Effect of cationic and anionic solid polymer electrolyte on direct electrochemical reduction of gaseous

$CO_2$ to fuel. Journal of $CO_2$ Utilization, 2013;3-4:49-55.

[90] Newman J, Thomas-Alyea KE. Electrochemical system. third edition, university of California, Berkeley, Inc publication, Copy write. TP255.N48. 2004.

[91] Velichenko A, Sarret M, Müller C. Nature of the passive film formed at a zinc anode in zinc–nickel containing solutions. Journal of Electroanalytical Chemistry. 1998;448:1-3.

[92] Qiao X, Li H, Zhao W, Li D. Effects of deposition temperature on electrodeposition of zinc–nickel alloy coatings. Electrochimical Acta. 2013;89:771-777.

[93] Baldwin K, Robinson M, Smith C. The corrosion resistance of electrodeposited zinc-nickel alloy coatings. Corrosion science. 1993;35:1267-1272.

[94] Alfantazi A, Brehaut G, Erb U. The effects of substrate material on the microstructure of pulse-plated Zn–Ni alloys. Surface and Coatings Technology. 1997;89:239-244.

[95] El Rehim SA, Fouad E, El Wahab SA, Hassan HH. Electroplating of zinc-nickel binary alloys from acetate baths. Electrochimical Acta. 1996;41:1413-1418.

Chapter 6

# C-H Activation/Functionalization via Metalla-Electrocatalysis

*Guilherme M. Martins, Najoua Sbei, Geórgia C. Zimmer and Nisar Ahmed*

## Abstract

In conventional methods, C–H activations are largely involved in the use of stoichiometric amounts of toxic and expensive metal & chemical oxidants, conceding the overall sustainable nature. Meanwhile, undesired byproducts are generated, that is problematic in the scale up process. However, electrochemical C–H activation via catalyst control strategy using metals as mediators (instead electrochemical substrate control strategy) has been identified as a more efficient strategy toward selective functionalizations. Thus, indirect electrolysis makes the potential range more pleasant, and less side reactions can occur. Herein, we summarize the metalla-electrocatalysis process for activations of inert C–H bonds and functionalization. These Metalla-electrocatalyzed C–H bond functionalizations are presented in term of C–C and C–X (X = O, N, P and halogens) bonds formation. The electrooxidative C–H transformations in the presence of metal catalysts are described by better chemoselectivities with broad tolerance of sensitive functionalities. Moreover, in the future to enhance sustainability and green chemistry concerns, integration of metalla-electrocatalysis with flow and photochemistry will enable safe and efficient scale-up and may even improve reaction times, kinetics and yields.

**Keywords:** metalla-electrocatalysis, C–H bonds activation, catalyst control strategy, mediators, atom and step economy

## 1. Introduction

The direct functionalization of C–H bonds provides a powerful synthetic pathway for selective C–C and carbon–heteroatom (C–X) bond formation, thus improving atom- and step economy as well as rationalization of chemical synthesis [1, 2]. In the field of conventional C–H activation, prefunctionalization of substrates, generally high temperatures, acidic conditions and/or the use of stoichiometric oxidants (such as a peroxide, a hypervalent iodine) are required due to the high bond dissociation energies, unreactive molecular orbital profiles, low acidities, and ubiquitous nature of the C–H bonds [3, 4]. The stoichiometric amount of reagents/oxidants affect the product's selectivity, additionally the formation of by-products result in overall low turnover of the reaction. Electrochemical C–H functionalization has advantages as this process avoids prefunctionalization of substrates and offering the direct transformation of a simple substrate to a complex and valuable molecule [5]. However, for C–H functionalization, still need a high oxidation potential for selective C–H bonds activation compared to organic solvents and common functional groups. To overcome this problem, indirect electrolysis via catalysts control strategy (mediators such as redox metal catalysts) is beneficial, makes the potential range more pleasant and has control over selectivity at mild conditions that is not observed through classical catalyst control strategy [6, 7]. From the 90's, great progress was observed in the evolution of reactions involving control of regioselectivity and enantioselectivity. Much of this merit was achieved by the evolution of catalysts based on high-performance transition metals. Derivatives of organic halides, triflates and several other leaving groups are still applied in reactions of aryl alkylations (Friedel-Crafts) and in cross-coupling reactions with several organometallic reagents. In addition, alkenes are also good substrates for aryl alkylation, alkenylation, or for cross-coupling reactions catalyzed by transition metals, using the corresponding halides or correlated substrates. However, most of the known transition metal catalysts do not meet all the requirements of modern developments, and often the biggest limitation is low efficiency and high costs to obtain efficient ligands. Faced with this challenge, there is an increasing use of new technologies applied concurrently to these catalytic systems, making transition metal catalysts more efficient and cleaner, enabling new mechanistic routes [8]. Additionally, the use of electrochemistry concomitant with the chemistry of transition metals offers a powerful strategy, since it avoids the use of external redox additives [9].

Electrocatalysis is a field of electrochemistry that has been gaining great growth in recent years due to the several advantages. In indirect electrosynthetic reactions, the exchange of electrons occurs between a mediator and the organic substrate. Therefore, by varying the applied current or voltage of the power source, the oxidation or reduction capacity of the electrochemical system can be manipulated, this being a great advantage in the method. Likewise, the redox mediator alters the applied potential required for electron transfer, making the potential range more pleasant, and fewer side reactions can occur, avoiding overoxidation, dimerization, parallel reactions or electrode passivation (**Figure 1**). In addition, electrocatalysis deals with the development for energy storage, solar fuels, fuel cells, and also other electrochemical devices with charge transfer reactions interfacial control [10].

Whereas the redox potentials and the selectivity of the reaction can be controlled by changing the ligand of the mediators of the complex transition metals. With the use of electrochemistry this process can become more selective, due to the possibility of controlling the electrical potential of the reaction

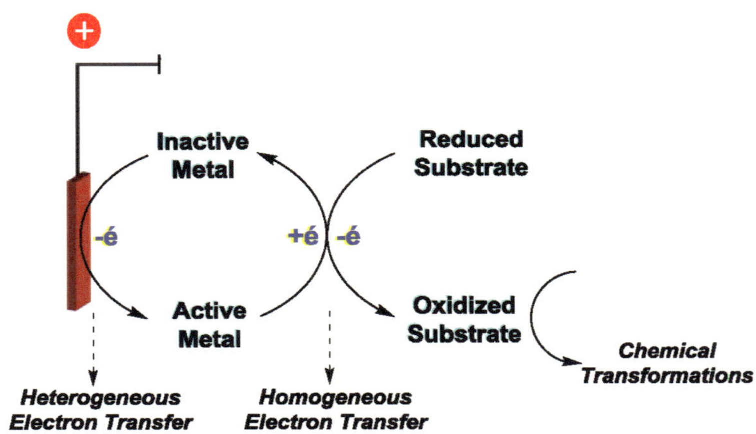


**Figure 1.**
*General illustration of indirect anodic transformation.*

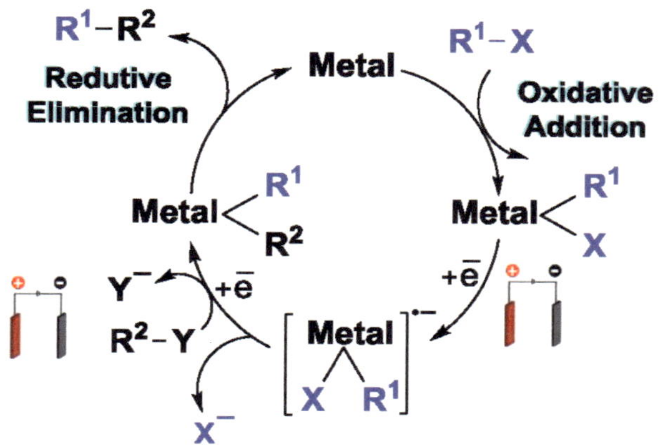


**Figure 2.**
*General mechanism for cross-coupling reactions with transition metal mediator.*

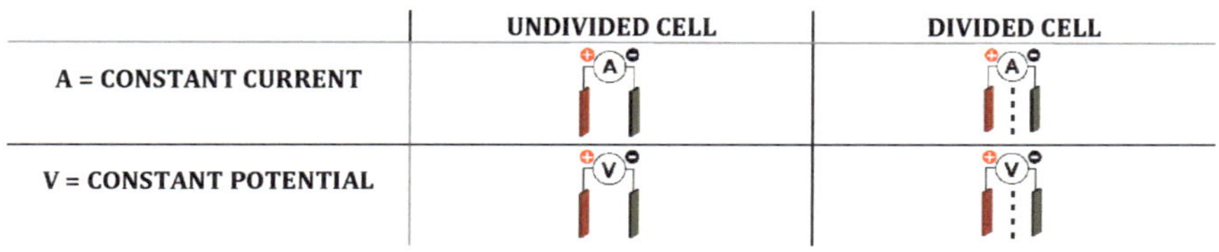


**Figure 3.**
*Cell notations.*

medium by changing the voltage (V) or the electric current (A) through the energy source (**Figure 2**) [11].

Considering this, we have prepared an overview of the recent metalla-electrocatalysis process for activation/functionalization of inert C–H bonds. The perspective and limitations together with mechanistic discussions will be presented.

To offer an easy interpretation of the different catalytic systems discussed here, we will use a standardized notation, differentiating the divided cell from the undivided cell, as well as if the reaction follows via constant current or constant potential (**Figure 3**). Additionally, the different types of electrodes will be added along with other details, offering a better experience for the reader.

## 2. C-C bond formation

The possibility of extending an organic structure through the formation of new C-C bonds is essential for medicinal chemistry, synthesis of natural products, materials chemistry and even agrochemical synthesis, among others [12–14]. Synthetic methodologies via carbometallation have been intensively developing in recent decades, and group **8–11** metals stand out in these transformations [15].

Palladium-catalyzed C-H cross-coupling reactions are known to be powerful tool to build new C-C bonds. Considering this, Mei and co-workers reported a C($sp^2$)-H coupling of ketoximes **1** with organoboron **2** or α-ketoacid **3** reagents catalyzed by Pd, using electrical current instead of external oxidants (**Figure 4**) [16]. In an H-type divided cell, with two platinum electrodes and a Nafion 117 membrane at 60°C,

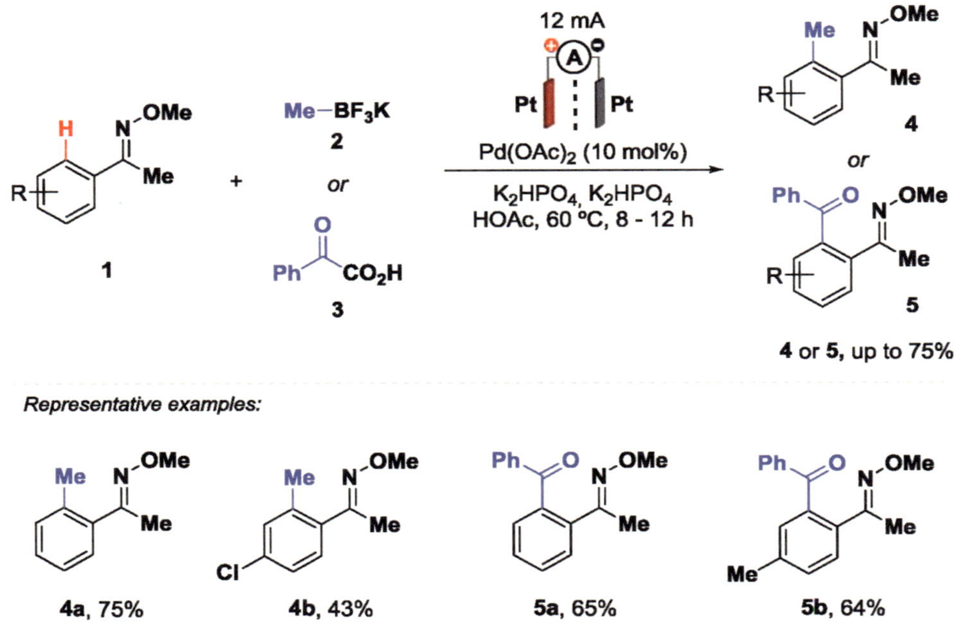


**Figure 4.**
*C(sp²)-H coupling catalyzed by Pd of ketoximes with organoboron or α-ketoacid reagents.*

several substituted oxime ethers were applied, and the corresponding methylated **4** and acylated **5** products were obtained, with yields of up to 75% isolated.

Considering experimental results, the authors suggest a mechanism for C(sp$^2$)-H methylation via electrochemical oxidation (**Figure 5**). Initially, the palladium catalyst coordinates with a nitrogen atom, approaching the ortho-C–H bond, activating the C(sp$^2$)-H bond to form the palladacycle. Transmetallation with $MeBF_3K$ under anodic oxidation conditions can provide Pd(III) or Pd(IV), which followed by reductive elimination, delivers the methylated product **4**, regenerating the Pd(II) species. The authors do not rule out the possibility of alkylation going via Pd(II)/Pd(0). It is worth mentioning that the cyclic voltammogram of palladacycle revealed an oxidation wave at 1.21 V *vs* Ag/AgCl, suggesting that the anode can oxidize the aryl palladium(II) intermediate to a high-valued Pd(III) or Pd(IV) species.

Asymmetric catalysis has valuable applications in the synthesis of useful compounds. A greater understanding of the mechanisms involved contributes to expanding its scope, as well as the use of new technologies, which should offer new insights. Ackermann and co-workers reported the very first asymmetric metalla-electrocatalyzed C-H activation [17]. With the aid of a transient directing group (TDG) using graphite felt and platinum electrodes, pallada-electrocatalysis was obtained in high enantioselectivities under moderate reaction conditions, providing the synthesis of highly enantiomerically-enriched biaryls axially chiral scaffolds **9** (**Figure 6**). Likewise, vinyl phosphonate, vinyl sulfone and cholesterol derivatives have increased the versatility of the method. Mechanistic experiments and computation studies provided important insights into the intermediates and the catalyst's path of action with the TDG. Kinetic studies with isotopically labeled substrates suggest that the activation of C-H is the determining step of the reaction.

C-centered radical cyclization under electrochemical conditions has been used to obtain cyclic structures. These radicals are highly reactive and attractive in organic synthesis, and has received attention. Pan and co-workers reported an electrosynthesis of functionalized 1-naphthols using alkynes and 1,3-dicarbonyl compounds by (4 + 2)

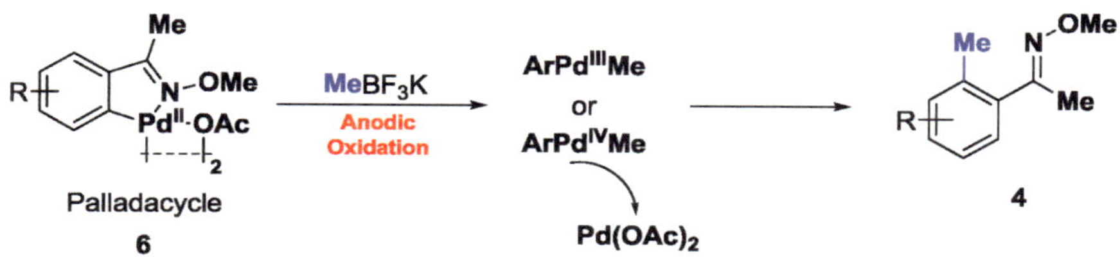


**Figure 5.**
*Representative mechanism for C(sp$^2$)-H coupling catalyzed by Pd of ketoximes with organoboron reagent.*

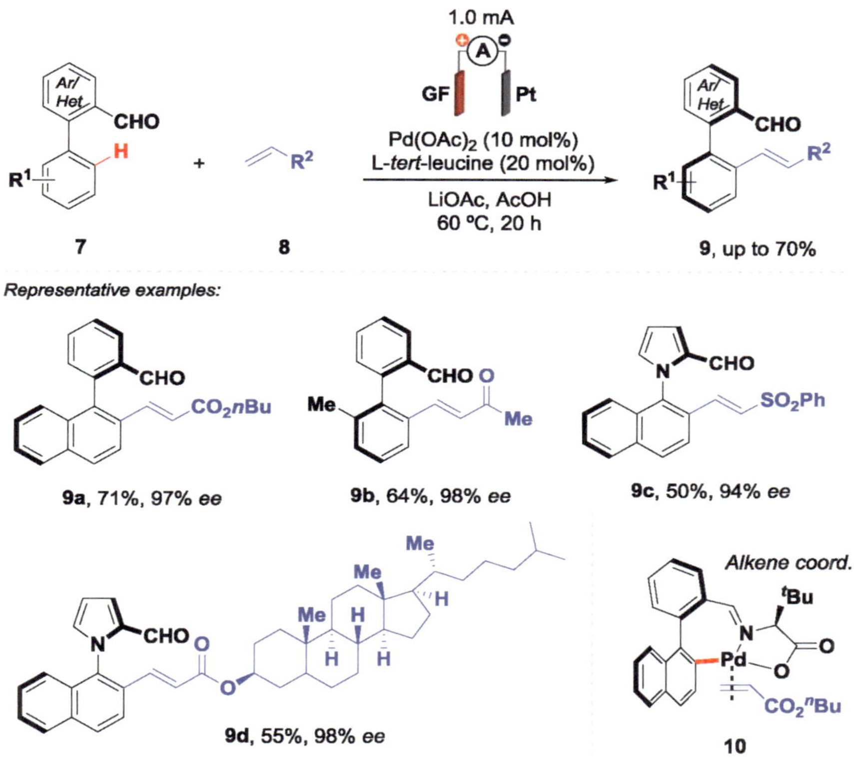


**Figure 6.**
*Asymmetric metalla-electrocatalyzed C-H activation for the synthesis of axially enantioenriched biaryl and heterobiaryl.*

annulation of C-centered radical [18]. The reactions were carried out in an undivided cell in the presence of $Cp_2Fe$ as a catalyst in THF/EtOH at a constant potential of 1.15 V vs. Ag/AgCl with NaOEt (30 mol%), during 2 h at 100°C (**Figure 7**). In general, good yield were obtained for compounds with the electron-donating or electron-withdrawing substituents, up to 84%. According to the control experiments, radical intermediates are involved and with absence of $Cp_2Fe$ the product was obtained with reduced yield, that is, direct electrolysis results in lower yields. Based on this and cyclic voltammetry experiments, a possible mechanism was proposed (**Figure 8**). Under electrochemical conditions, it is necessary to form the conjugate base **14** to react with $Cp_2Fe$ due the oxidation potential of intermediate **14** is slightly lower than $Cp_2Fe$. The ethoxy ion was formed from cathodic reduction and reacts with the compound **1a** to form the intermediate **14**. Meanwhile, at the anode, $Cp_2Fe$ is oxidized to $Cp_2Fe^+$, which can be oxidized to intermediate **14** to conduct the single-electron transfer, generating a C-radical intermediate **15**. The radical intermediate **15** react with compound **12** to give

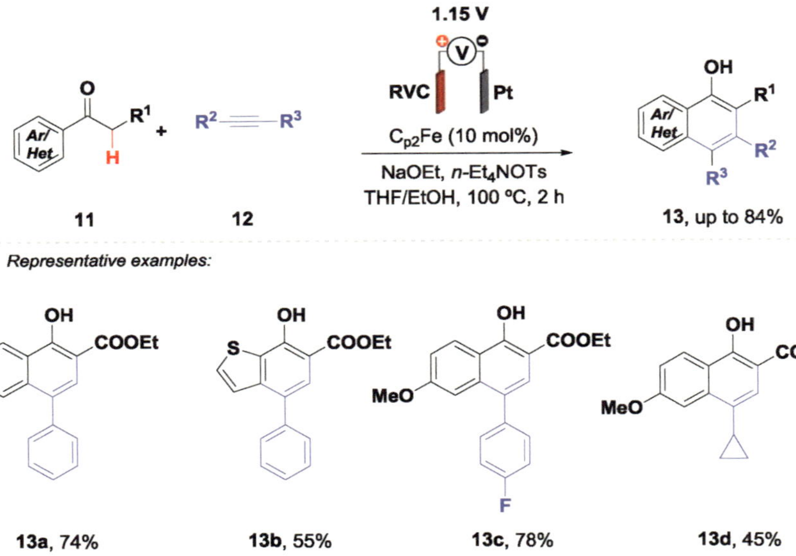


**Figure 7.**
*Electrochemical intermolecular annulation of alkyne with 1,3-dicarbonyl.*

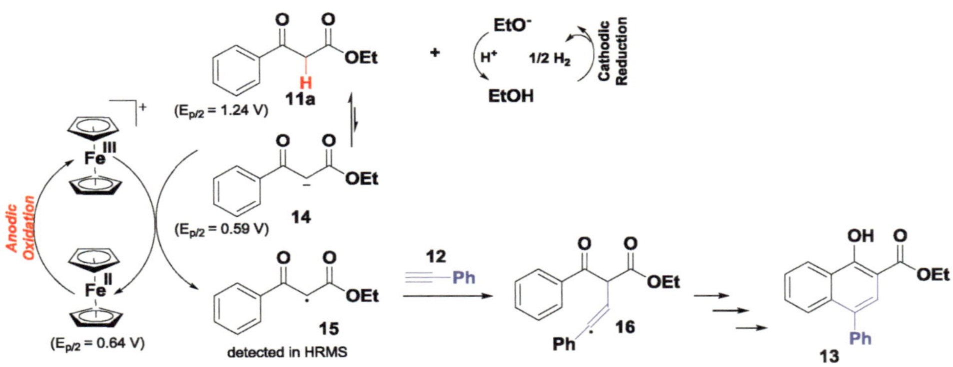


**Figure 8.**
*Proposed mechanism for electrosynthesis of functionalized 1-naphtols.*

intermediate **16**. From this point an intramolecular cyclization occur leading to obtain the product **13**.

Ackermann and co-workers reported an electrooxidative C-C alkenylation performed by rhodium(III) catalysis [19]. This reaction proceeded with ample scope and excellent levels of chemo- and position selectivities within an organometallic C-C activation manifold. The reactions were carried out in an undivided cell, in a constant current at 4.0 mA using $[Cp^*RhCl_2]_2$ as catalyst, in combination with a platinum plate cathode and a reticulated vitreous carbon (RVC) at anode, along with KOAc as additive in $H_2O$ at 100°C (**Figure 9**). According to the examination of leaving group substitution pattern, tertiary and secondary alcohols bearing either aryl or alkyl groups led the product **19**. Contrarily, a primary alcohol did not deliver the desired product, illustrating the importance of the acidic functionality for inducing the C-C cleavage. This methodology revealed to be a position-selective rhodium-catalyzed C-C activation of 1,2,3-trisubstituted arenes **17**. Mechanism analysis showed that C-C activation occurred significantly faster as compared to corresponding C–H activation. Furthermore, the presence of molecular hydrogen

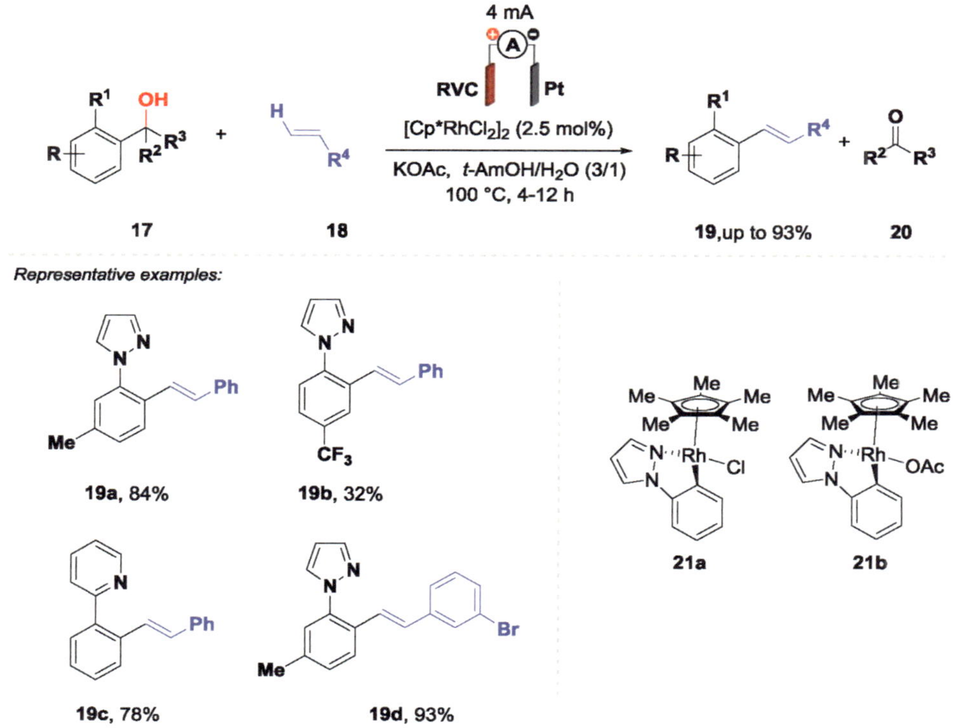


**Figure 9.**
*Eletrochemical C–C alkenylation by rhodium(III) catalysis.*

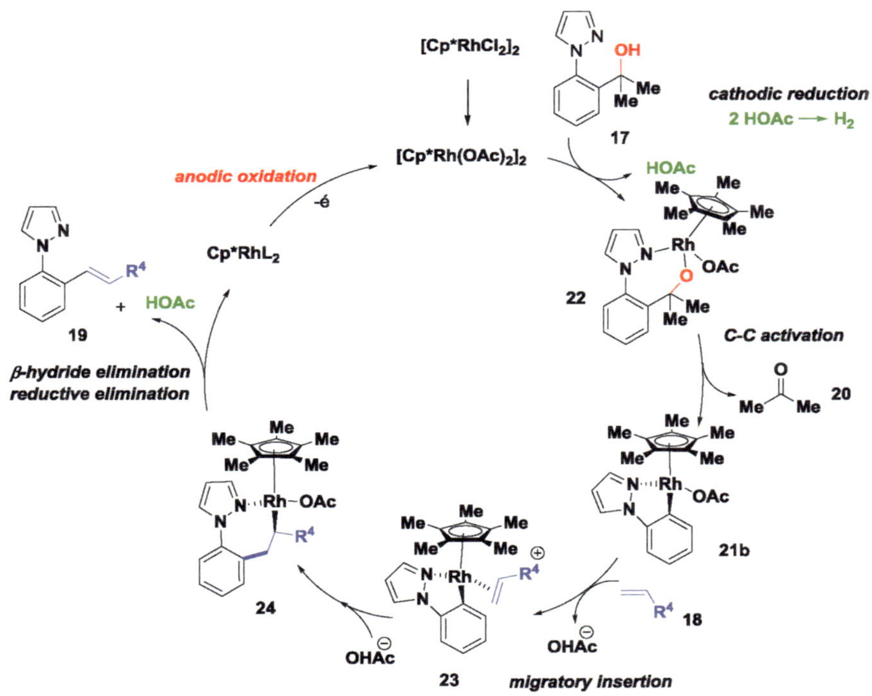


**Figure 10.**
*Proposed catalytic cycle for the rhodium-electrocatalysed C–C alkenylation.*

as byproduct was confirmed by gas-chromatographic headspace analysis. The previously prepared complex **21a-b** showed to be a competent catalyst, proving the organometallic nature of the electro-oxidative C–C alkenylation. The cyclic voltammetry experiments showed clearly a ligand exchange, forming $[Cp^*Rh(OAc)_2]_2$. The proposed reoxidation of rhodium(I) species to regenerate the catalytically competent rhodium(III) was explored with the well-defined Cp*Rh(I) complex [Cp*Rh(cod)]. This complex was shown to be easily oxidized at Ep = −0.16 V versus $Fc^{+/0}$. Based on this study a plausible catalytic cycle for the rhodium-electrocatalyzed C-C alkenylation was proposed (**Figure 10**).

Mo and co-workers reported a general electrochemical strategy for the combined trifluoromethylation/$C(sp^2)$–H functionalization using Langlois' reagent as the $CF_3$ source [20]. The reactions were carried out an undivided cell using $MnBr_2$ as the mediator, $H_3PO_4$ as the sacrificial oxidant, Pt as the electrodes with a constant electric current of 10 mA for 6 h (**Figure 11**).

The mechanism study by cyclic voltammetry showed that combination of $MnBr_2$ and $CF_3SO_2Na$ exhibits a quasi-reversible anodic CV feature at 0.83 V, that was attributed to the $Mn^{II}/Mn^{III}$ redox couple of the $CF_3$-bond complex. When the reagent was added in mixture of $MnBr_2$ and $CF_3SO_2Na$, it was observed the presence of two irreversible anodic waves of 0.94 and 1.59 V, which correspond to the formation of the putative $Mn^{III}$-$CF_3$ and the single electron oxidation leading to the final product. Summarizing, the $Mn^{III}$-$CF_3$ species is produced by anodic oxidation of $Mn^{II}$ in the presence of Langlois reagent. After, Mn-assisted delivery of $CF_3$• to the olefin forming a carbon radical. Subsequently, the aromatic ring radical is

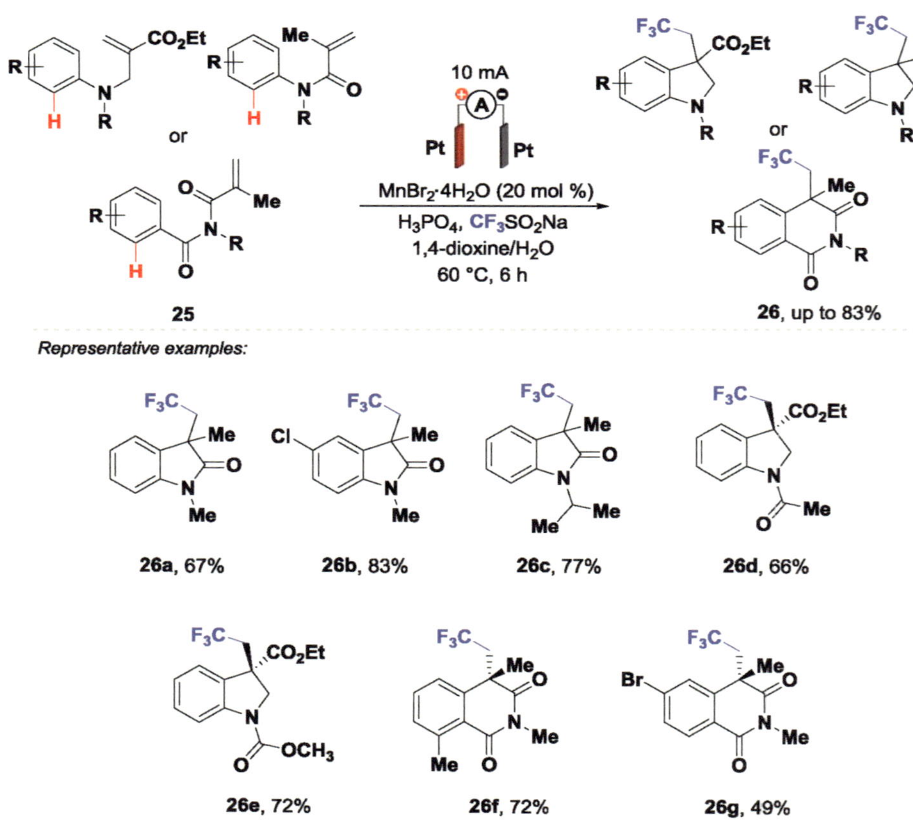


**Figure 11.**
*Oxidant-free electrochemical trifluoromethylation-initiated radical oxidative cyclization.*

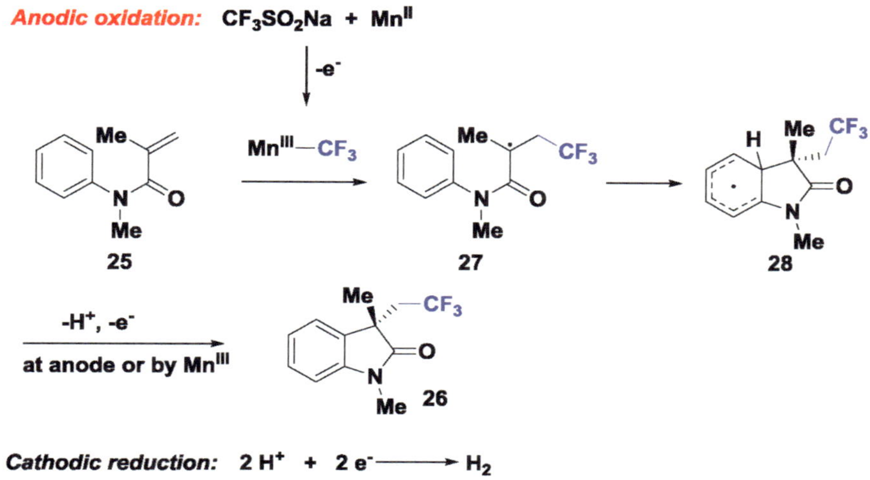


**Figure 12.**
*Proposed mechanism for trifluoromethylation/C(sp²)–H functionalization.*

formed and following by either anode or $Mn^{III}$-mediated oxidation, then product **26** was obtained (**Figure 12**).

## 3. C-X (X = O, N, P, halogen) bond formation

The organic molecules with C-N, C-P, C-O, and C-Cl bond play an important role in the biological application, such as drug synthesis, agrochemicals, etc. [21–24]. Therefore, the new synthetic strategies to form a carbon-heteroatom bond have been made in the developments of various electrochemical methods based on metal-catalysis such as Pd, Co, Mn, Ag, and Rh. In this context, Lei and co-workers reported a C-H/N-H coupling catalyzed by Pd to synthesize of pyrido[1,2-*a*] benzimidazole [25]. Under the mild condition, different N-phenylpyridin-2-amine could afford the desired product in yields of up to 99% (**Figure 13**). The reaction was performed in an undivided cell equipped with a carbon plate as anode and a Fe plate as a cathode, under a constant current, using the system $CH_3CN/LiClO_4$ as a solvent/electrolyte.

As an improvement of this transformation, the authors suggest a mechanism for C-H/N-H coupling reaction catalyzed by Pd(II) via electrochemical oxidation (**Figure 14**). Initially, Pd(II) coordinates with a nitrogen atom of substrate **29** to form the intermediate **31**, which gives the complex intermediate **32** after electrophilic deprotonation. The latter then underwent a reductive elimination process to provide the desired product **30** and Pd(0). Finally, Pd(0) oxidized at the anode to be recovered to Pd(II).

Cobalt-catalyzed C-H cross-coupling reactions are known to be a strong implement to build new C-N bonds [26]. In this context, Lei and co-workers reported a C(sp²)-H coupling catalyzed by Co of quinoline amide **32** with secondary amine **33** (**Figure 15**) [27]. Under a constant current of 10 mA, a large family of desired product **34** was obtained in moderate to good yields up to 74%. The reaction proceeds in a divided cell equipped with a carbon plate as an anode in acetonitrile and a Ni plate cathode in methanol. Independently from Lei group, Ackermann group [28] also reported the Co-catalyzed electrooxidative reaction of amides derivatives **35** and a secondary amine **36** (**Figure 16**). The authors achieved the best results in an undivided cell equipped with an RVC and Pt as the anode/cathode system,

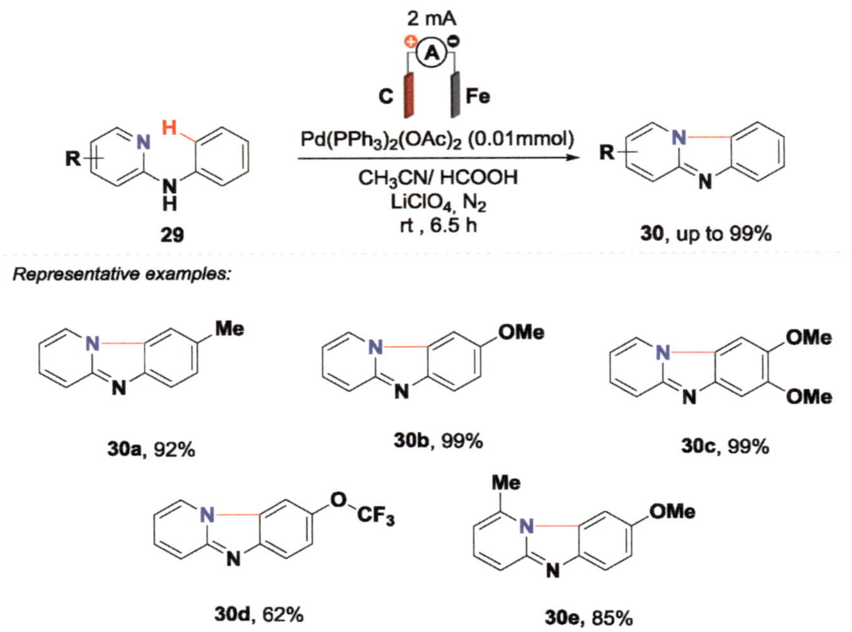


**Figure 13.**
*Pd-catalyzed C-N bond formation.*

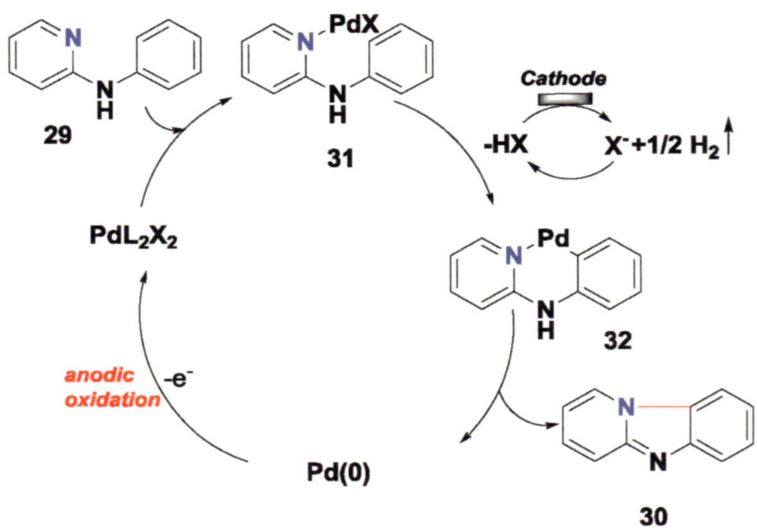


**Figure 14.**
*A plausible mechanism for C($sp^2$)-H coupling catalyzed by Pd of N-phenylpyridin-2-amine.*

at a constant current of 2.5 mA. The desired products **37** were formed in excellent yields of up to 83%.

As an improvement of this methodology, the authors suggest a plausible mechanism (**Figure 17**). In the path I: Co(II) is oxidized at the anode to give Co(III); which coordinates with N-(quinolin-8-yl)benzamide **35** to form Co(III)-species **39**. In the Path II: Co(II) coordinated to N-(quinolin-8-yl)benzamide **35** to get Co(II)-complex **38**, in the presence of a base. This Co(II)- species **38** is oxidized at the anode to provide Co(III)-species **39**. Then, C–H activation took place by the base,

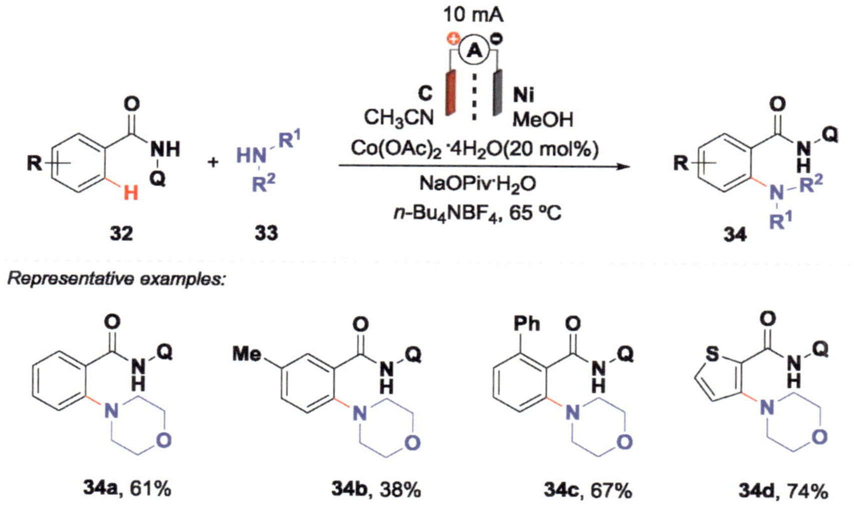

**Figure 15.**
*Co-catalyzed C-N bond formation.*

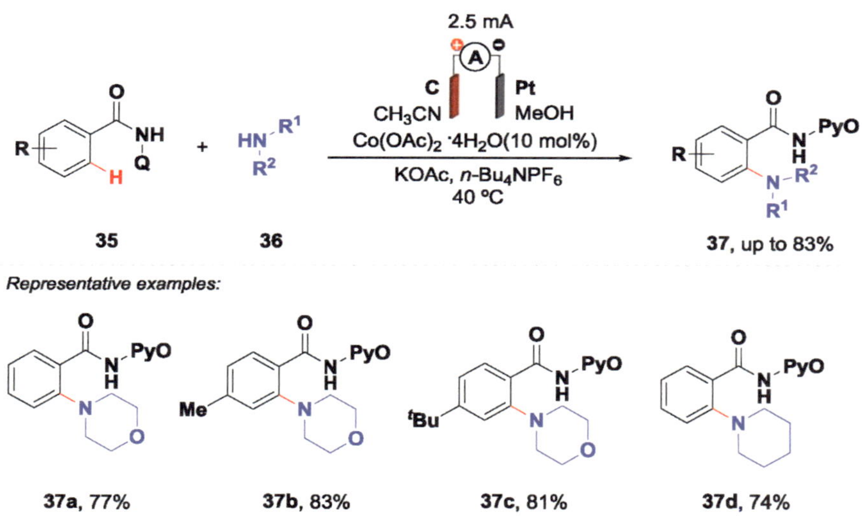

**Figure 16.**
*Co-catalyzed C-N bond formation.*

and Co(III)-species **39** was attacked by **36** to form Co(III)-species **40**, followed by reductive elimination of Co(III)-complex **40** to release the desired product and Co(I) species. Finally, Co(I) species was reoxidized to Co(II) at the anode to complete the whole catalytic cycle of Co.

The Mn-catalyzed formation of the C-Cl bond was reported by Chen and co-workers (**Figure 18**) [29]. Electrolyzing styrene derivatives **42** in the presence of $O_2$ gas and $MgCl_2$ at a constant current afford a large family of desired products **43** in very good yields. The reaction proceeds in an undivided cell equipped with a carbon rod both as anode and cathode, using the system Acetone-DCM/$LiClO_4$ as a solvent/electrolyte, for 12 h (**Figure 18**).

A mechanistic elucidation in **Figure 19** shows that first, Mn(II)Cl oxidized at the anode providing Mn(III)Cl species. Then styrene derivatives **42** reacts with Mn(III)

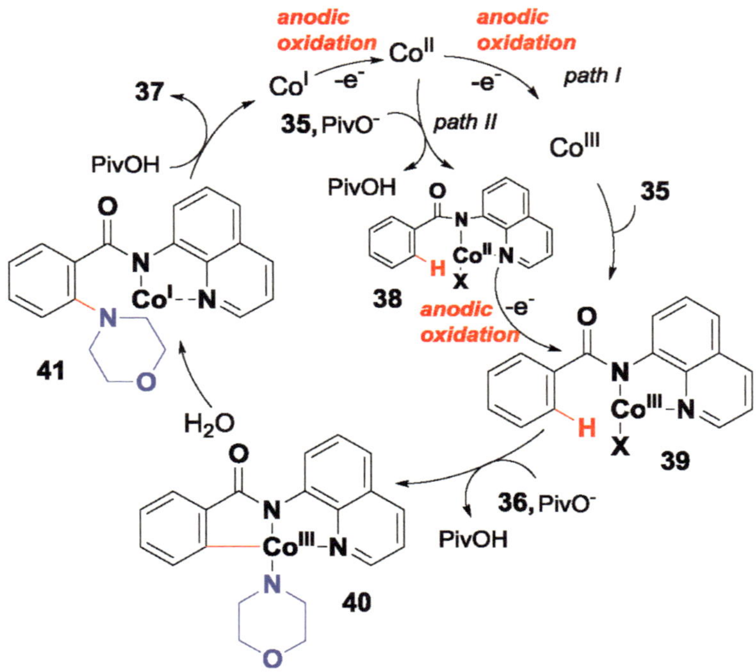


**Figure 17.**
*A plausible mechanism for C-N bond formation by Co-catalysis.*

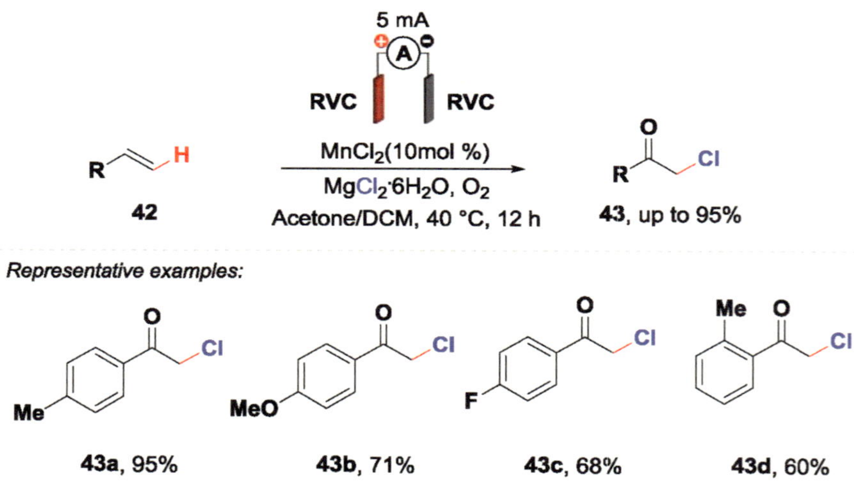


**Figure 18.**
*Mn-catalyzed C-Cl bond formation.*

Cl to provide intermediate **44**. At the same time, at the cathode, the reduction of $O_2$ gives the radical superoxide ion which easily reacts with **44** to generate intermediate **45**. This later decomposes to form compound **46**. After further oxidation of **46**, the desired products **43** was formed.

Budnikova and co-workers reported an efficient approach of Ag-catalyzed reaction to a range azole dialkyl phosphonates derivatives **49** [30]. Under mild conditions, different substituted azole **47** and dialkyl-*H*-phosphonates **48** afford the final products **49** in moderate to good yields up to 75% (**Figure 20**).

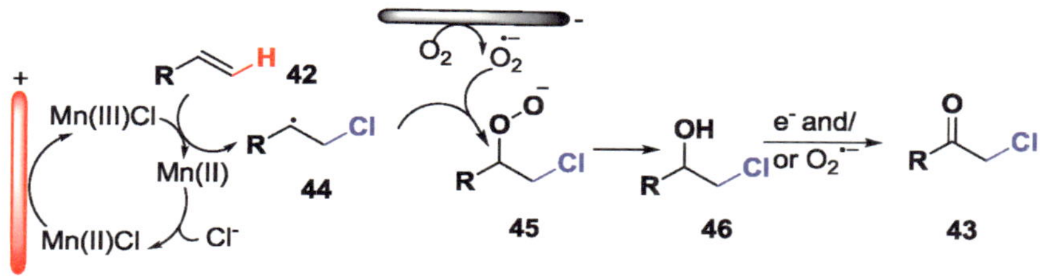


**Figure 19.**
*A plausible mechanism for C-Cl bond formation by Mn-Catalysis.*

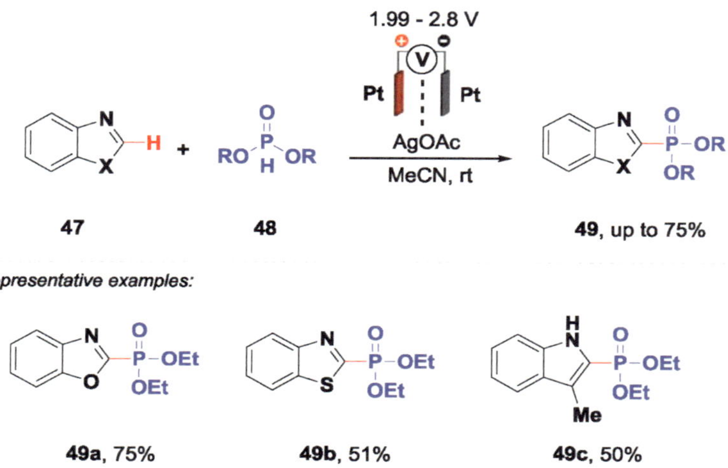


**Figure 20.**
*Ag-catalyzed C-P bond formation.*

The electrolysis proceeds in a divided cell at a constant voltage, employing AgOAc and $Na_3PO_4$ as additives and using acetonitrile as solvent. The proposed mechanism of this methodology is described in **Figure 21**. The reaction starts by combining dialkyl-*H*-phosphonate and silver (I) cation leading intermediate **50**, which after oxidation gives the radical intermediate **51**. Then azole derivatives **47** coordinate with **51** to form radical **52**. This latter, after losing hydrogen cation and an electron, leads to the desired product **49**.

Xu and co-workers reported an efficient method for rhodium (III)-electrocatalyzed to form the C-P bond (**Figure 22**) [31]. Using a graphite rod as anode and a platinum plates as a cathode, different substituted *N*-(2-pyridyl) aniline **53** and phosphine oxide **54** could provide the final product **55** in high yields. The electrolysis was performed in an undivided cell, under reflux in methanol at a constant current.

A possible mechanism of this strategy is shown in **Figure 23**. The reaction starts with C-H activation in phenylpyridine **53** by the catalyst **56** to give intermediate **57**. A further insertion of diphenylphosphine oxide **54** gives intermediate **58**. This later undergoes anodic oxidation forming to products **55**, regenerating the active complex **56**.

Strekalova and co-workers developed an elegant approach for Co-catalysed electrochemical formation of the C-P bond [32]. By using cobalt complex as a catalyst, different diethyl phosphonates **61** and aryl derivatives **60** could afford the desired products **62** with yields up to 80% for reductive condition and up to 68%

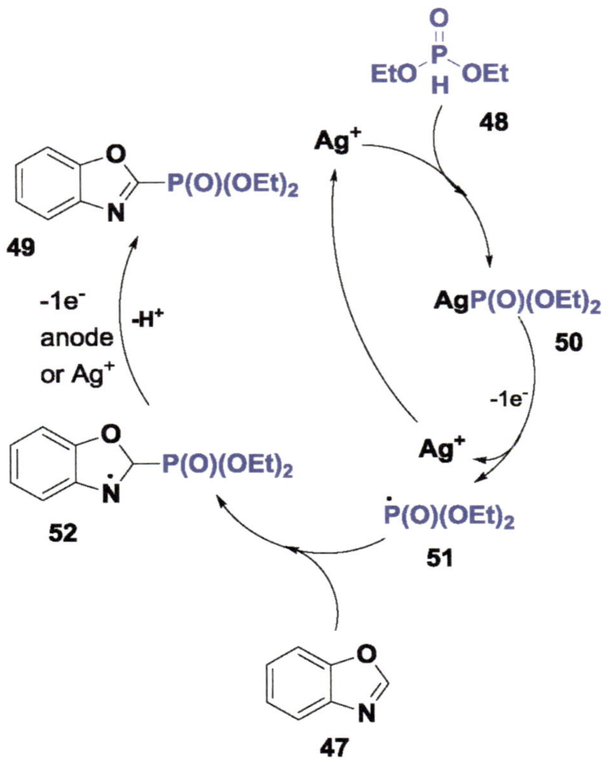


**Figure 21.**
*A plausible mechanism for C-P bond formation by Ag-Catalysis.*

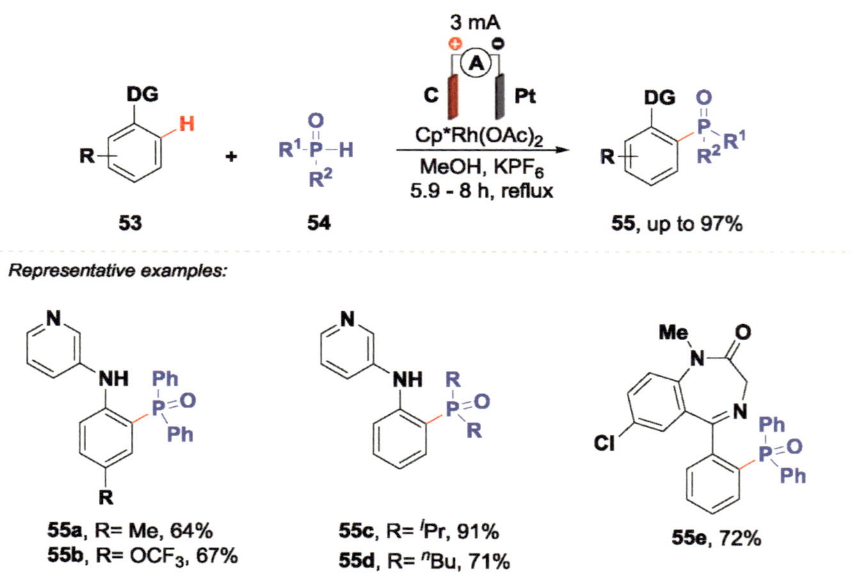


**Figure 22.**
*Rh(III)- catalyzed electrochemical phosphorylation of aryl substrates.*

for oxidative conditions. The electrolysis was carried out under a constant voltage of −0.3 V vs. $Fc^{+}/Fc$ in a divided cell, equipped with platinum electrodes both as anode and cathode (**Figure 24**).

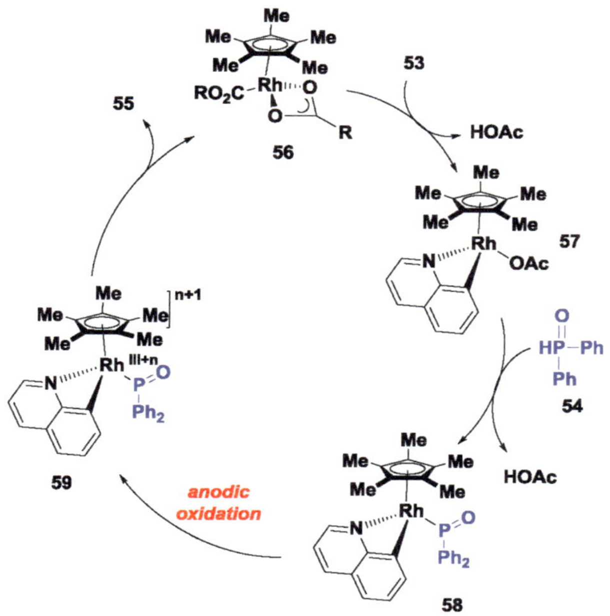


**Figure 23.**
*Plausible mechanism for Rh(III)- catalyzed electrochemical phosphorylation of aryl substrates.*

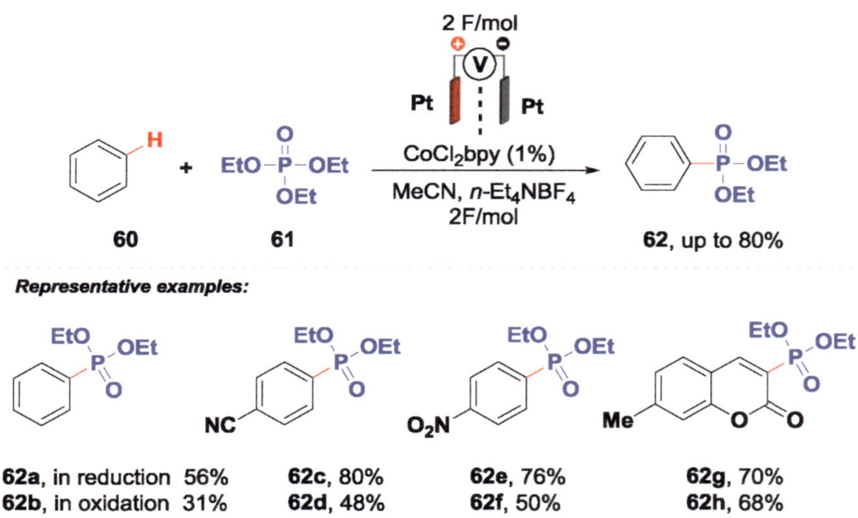


**Figure 24.**
*Co-catalyzed C-P bond formation.*

The plausible mechanism (**Figure 25**) shows that at the start, $Co^{2+}$ precursors coordinates with H-phosphonate **61** to give complex intermediate **63**, which after further oxidation (or reduction), leads the intermediate **64** (or **65**). **B** (or **C**) forms after proton elimination a radical intermediate **67**. Then, the insertion of **60** provides the final products **62**.

The C-O bond formation under Co-catalyst was reported by Ackermann group (**Figure 26**) [33]. A variety of amides **68** and primary alcohols **69** were electrolyzing at a constant current of 8 mA as a green oxidant in a simple undivided cell equipped with carbon as anode and a platinum cathode for 6 h, providing the desired product **70** with good yields.

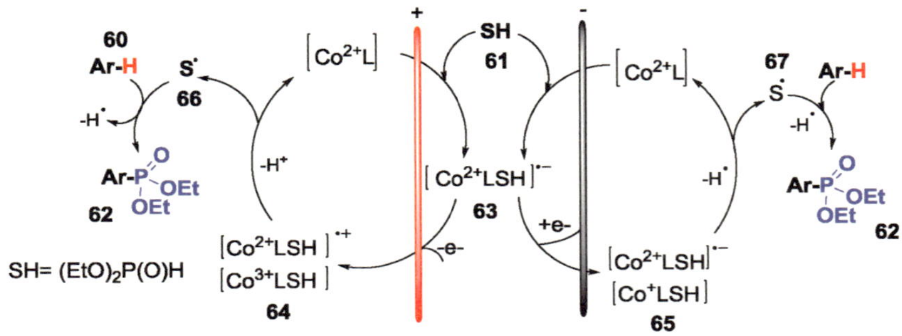


**Figure 25.**
*A plausible mechanism for C-P bond formation by Co-Catalysis.*

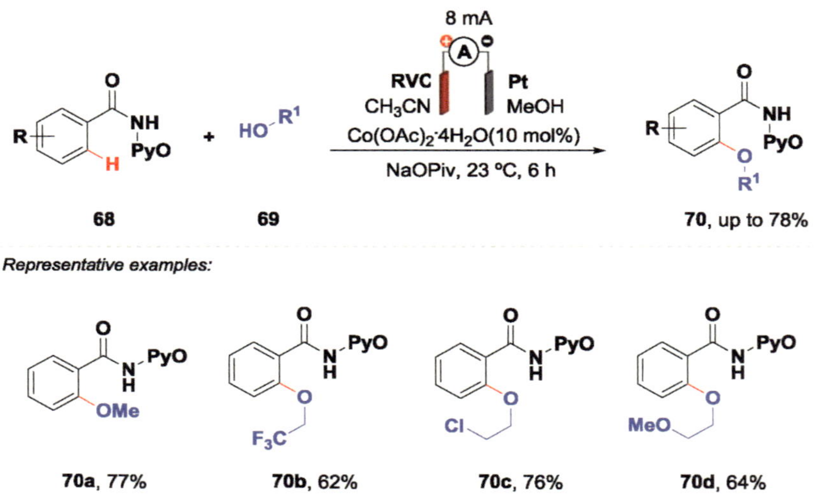


**Figure 26.**
*Co-catalyzed C-O bond formation.*

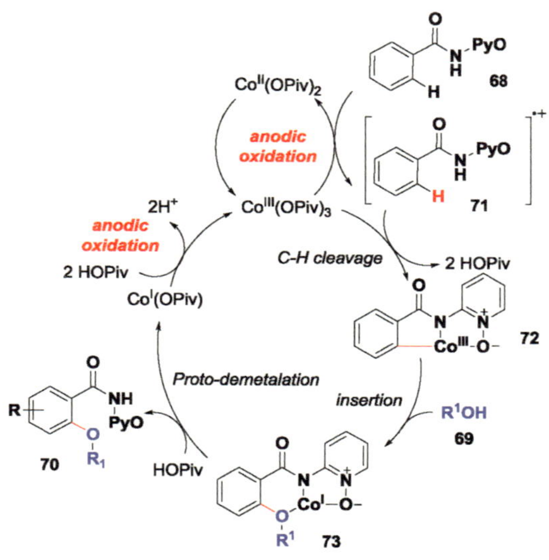


**Figure 27.**
*A plausible mechanism for C-O bond formation by Co-Catalysis.*

Presumably, a catalytic cycle commences with the oxidation of $Co^{II}$ precatalyst at the anode to give a $Co^{III}$ species capable of forming complex **72**. Successive addition of alcohol derivatives leads to complex **73**, which in the presence of HOPiv gives the final product and forms a $Co^{I}$ species. The latter, which is oxidized at the anode, gives a catalytically active $Co^{III}$ species (**Figure 27**).

## 4. Conclusions

C–H activation/functionalization via metalla-electrocatalysis appears as a valuable tool for organic synthesis. Coupling reactions with hydrogen evolution demonstrate great potential for application in the synthesis of complex molecules. Likewise, electrochemical C–H activation appears to be a greener method, and even more progress is expected in this area of research. However, there are still several challenges, such as the application of other transition metals, the recycling of transition metals and electrolytes, an in-depth study of asymmetric transformations, the application of new ligands, etc. Despite all these challenges, we believe that in the future, the fusion between the transition metal catalysis with electrochemical methods will have a great development, being highly promising for synthetic chemistry, becoming a common tool in all research laboratories.

## Acknowledgements

Support from the EU-COFUND project (Grant No 663830) to Dr. Nisar Ahmed is gratefully acknowledged. We also thank the School of Chemistry, Cardiff Chemistry (N.A), CNPq (G. M. M and G. C. Z) and RUDN University 5–100 Project framework Program (N.S) for their support.

## Conflict of interest

The authors declare no conflict of interest.

## Author details

Guilherme M. Martins[1,2†], Najoua Sbei[3†], Geórgia C. Zimmer[2†] and Nisar Ahmed[1*]

1 School of Chemistry, Cardiff University, Main Building Park Place, Cardiff, CF10 3AT, United Kingdom

2 Chemistry Department, Federal University of Santa Catarina, Florianópolis, 88040-900, SC, Brazil

3 Organic Chemistry Department, Peoples' Friendship University of Russia (RUDN University), 6 Miklukho-Maklaya St., Moscow, 117198, Russian Federation

*Address all correspondence to: ahmedN14@cardiff.ac.uk

† These authors contributed equally to this work.

## References

[1] Yang K, Song M, Liu H, Ge H. Palladium-catalyzed direct asymmetric C–H bond functionalization enabled by the directing group strategy. Chem Sci. 2020; DOI: 10.1039/D0SC03052J

[2] Yamaguchi J, Yamaguchi AD, Itami K. C-H Bond Functionalization: Emerging Synthetic Tools for Natural Products and Pharmaceuticals. Angew Chemie Int Ed. 2012;51(36):8960-9009. DOI: 10.1002/anie.201201666

[3] Santoro S, Kozhushkov SI, Ackermann L, Vaccaro L. Heterogeneous catalytic approaches in C–H activation reactions. Green Chem. 2016;18(12):3471-3493. DOI: 10.1039/C6GC00385K

[4] Blanksby SJ, Ellison GB. Bond Dissociation Energies of Organic Molecules. Acc Chem Res. 2003;36(4):255-263. DOI: 10.1021/ar020230d

[5] Martins GM, Zimmer GC, Mendes SR, Ahmed N. Electrifying green synthesis: recent advances in electrochemical annulation reactions. Green Chem. 2020;22(15):4849-4870. DOI: 10.1039/D0GC01324B

[6] Martins GM, Shirinfar B, Hardwick T, Murtaza A, Ahmed N. Organic electrosynthesis: electrochemical alkyne functionalization. Catal Sci Technol. 2019;9(21):5868-5881. DOI: 10.1039/C9CY01312A

[7] Margarita C, Lundberg H. Recent Advances in Asymmetric Catalytic Electrosynthesis. Catalysts. 2020;10(9):982. DOI: 10.3390/catal10090982

[8] Cheng W-M, Shang R. Transition Metal-Catalyzed Organic Reactions under Visible Light: Recent Developments and Future Perspectives. ACS Catal. 2020;10(16):9170-9196. DOI: 10.1021/acscatal.0c01979

[9] Chen J, Lv S, Tian S. Electrochemical Transition-Metal-Catalyzed C−H Bond Functionalization: Electricity as Clean Surrogates of Chemical Oxidants. ChemSusChem. 2019;12(1):115-132. DOI: 10.1002/cssc.201801946

[10] Koper MTM, Iwasawa Y. Electrocatalysis. Phys Chem Chem Phys. 2014;16(27):13567.

[11] Fuchigami T, Atobe M, Inagi S. Fundamentals and Applications of Organic Electrochemistry [Internet]. Fuchigami T, Inagi S, Atobe M, editors. Chichester, United Kingdom: John Wiley & Sons Ltd; 2014. DOI: 10.1002/9781118670750

[12] Ma D, Wang Y, Liu A, Li S, Lu C, Chen C. Covalent Organic Frameworks: Promising Materials as Heterogeneous Catalysts for C-C Bond Formations. Catalysts. 2018;8(9):404. DOI: 10.3390/catal8090404

[13] Hong B-C, Nimje R. Catalytic C-C Bond Formation in Natural Products Synthesis: Highlights From The Years 2000 – 2005. Curr Org Chem. 2006;10(17):2191-2225. DOI: 10.2174/138527206778742605

[14] Legros J, Figadère B. Iron-promoted C–C bond formation in the total synthesis of natural products and drugs. Nat Prod Rep. 2015;32(11):1541-1555. DOI: 10.1039/C5NP00059A

[15] Fensterbank L, Goddard J-P, Malacria M. C–C Bond Formation (Part 1) by Addition Reactions: through Carbometallation Catalyzed by Group 8-11 Metals. In: Comprehensive Organometallic Chemistry III. Elsevier; 2007. p. 299-368. DOI: 10.1016/B0-08-045047-4/00129-1

[16] Ma C, Zhao C-Q, Li Y-Q, Zhang L-P, Xu X-T, Zhang K, et al. Palladium-catalyzed C–H activation/C–C cross-coupling reactions via electrochemistry. Chem Commun. 2017;53(90):12189-12192. DOI: 10.1039/C7CC07429H

[17] Dhawa U, Tian C, Wdowik T, Oliveira JCA, Hao J, Ackermann L. Enantioselective Pallada-Electrocatalyzed C−H Activation by Transient Directing Groups: Expedient Access to Helicenes. Angew Chemie Int Ed. 2020;59(32):13451-13457. DOI: 10.1002/anie.202003826

[18] He M-X, Mo Z-Y, Wang Z-Q, Cheng S-Y, Xie R-R, Tang H-T, et al. Electrochemical Synthesis of 1-Naphthols by Intermolecular Annulation of Alkynes with 1,3-Dicarbonyl Compounds. Org Lett. 2020;22(2):724-728. DOI: 10.1021/acs.orglett.9b04549

[19] Qiu Y, Scheremetjew A, Ackermann L. Electro-Oxidative C–C Alkenylation by Rhodium(III) Catalysis. J Am Chem Soc. 2019;141(6):2731-2738.

[20] Zhang Z, Zhang L, Cao Y, Li F, Bai G, Liu G, et al. Mn-Mediated Electrochemical Trifluoromethylation/ C($sp_2$)–H Functionalization Cascade for the Synthesis of Azaheterocycles. Org Lett. 2019;21(3):762-766. DOI: 10.1021/acs.orglett.8b04010

[21] Evano G, Blanchard N, Toumi M. Copper-Mediated Coupling Reactions and Their Applications in Natural Products and Designed Biomolecules Synthesis. Chem Rev. 2008;108(8):3054-3131.

[22] MATSUMOTO Y, UCHIDA W, NAKAHARA H, YANAGISAWA I, SHIBANUMA T, NOHIRA H. Novel Potassium Channel Activators. III. Synthesis and Pharmacological Evaluation of 3,4-Dihydro-2H-1,4-benzoxazine Derivatives: Modification at the 2 Position. Chem Pharm Bull (Tokyo). 2000;48(3):428-432.

[23] Hidaka T, Imai S, Hara O, Anzai H, Murakami T, Nagaoka K, et al. Carboxyphosphonoenolpyruvate phosphonomutase, a novel enzyme catalyzing C-P bond formation. J Bacteriol. 1990;172(6):3066-3072.

[24] Kam C-M, Kerrigan JE, Plaskon RR, Duffy EJ, Lollar P, Suddath FL, et al. Mechanism-Based Isocoumarin Inhibitors for Blood Coagulation Serine Proteases. Effect of the 7-Substituent in 7-Amino-4-chloro-3-(isothioureidoalkoxy)isocoumarins on Inhibitory and Anticoagulant Potency. J Med Chem. 1994;37(9):1298-1306.

[25] Duan Z, Zhang L, Zhang W, Lu L, Zeng L, Shi R, et al. Palladium-Catalyzed Electro-oxidative C–H Amination toward the Synthesis of Pyrido[1,2- a ]benzimidazoles with Hydrogen Evolution. ACS Catal. 2020;10(6):3828-3831.

[26] Mei R, Koeller J, Ackermann L. Electrochemical ruthenium-catalyzed alkyne annulations by C–H/Het–H activation of aryl carbamates or phenols in protic media. Chem Commun. 2018;54(91):12879-12882. DOI: 10.1039/C8CC07732K

[27] Gao X, Wang P, Zeng L, Tang S, Lei A. Cobalt(II)-Catalyzed Electrooxidative C–H Amination of Arenes with Alkylamines. J Am Chem Soc. 2018;140(12):4195-4199.

[28] Sauermann N, Mei R, Ackermann L. Electrochemical C−H Amination by Cobalt Catalysis in a Renewable Solvent. Angew Chemie Int Ed. 2018;57(18):5090-5094.

[29] Tian S, Jia X, Wang L, Li B, Liu S, Ma L, et al. The Mn-catalyzed paired electrochemical facile oxychlorination of styrenes via the oxygen reduction

reaction. Chem Commun. 2019;55(80):12104-12107.

[30] Yurko EO, Gryaznova T V., Kholin K V., Khrizanforova V V., Budnikova YH. External oxidant-free cross-coupling: electrochemically induced aromatic C–H phosphonation of azoles with dialkyl- H -phosphonates under silver catalysis. Dalt Trans. 2018;47(1):190-196. DOI: 10.1039/C7DT03650G

[31] Wu Z, Su F, Lin W, Song J, Wen T, Zhang H, et al. Scalable Rhodium(III)-Catalyzed Aryl C–H Phosphorylation Enabled by Anodic Oxidation Induced Reductive Elimination. Angew Chemie Int Ed. 2019;58(47):16770-4.

[32] StrekalovaSO,GrinenkoVV.,Gryaznova T V., Kononov AI, Dolengovski EL, Budnikova YH. Electrochemical phosphorylation of arenes catalyzed by cobalt under oxidative and reductive conditions. Phosphorus, Sulfur Silicon Relat Elem. 2019;194(4-6):506-509.

[33] Sauermann N, Meyer TH, Tian C, Ackermann L. Electrochemical Cobalt-Catalyzed C–H Oxygenation at Room Temperature. J Am Chem Soc. 2017;139(51):18452-18455.

Chapter 7

# Recent Trends in Development of Metal Nitride Nanocatalysts for Water Electrolysis Application

*Akhoury Sudhir Kumar Sinha and Umaprasana Ojha*


## Abstract

Nanocatalysts for sustainable water electrolysis is strongly desirable to promote the commercialization of $H_2$ as the alternate clean energy source for the future. The goal is cheaper hydrogen production from sea and low grade water by minimizing the energy consumption and using low cost cell components & non-noble metal catalysts. The conductivity of metal nitrides and their ability to carry out Hydrogen Evolution Reaction and Oxygen Evolution Reaction at relatively low overpotential render these one of the frontline candidates to be potentially utilized as the catalyst for low cost $H_2$ production via electrolysis. In this chapter, the potential of metal nitride catalyst towards fulfilling the above objective is discussed. The synthesis of various metal nitride catalysts, their efficiency towards electrode half reactions and the effectiveness of these class of nanocatalyst for electrolysis of sea water is elaborated. A review of recent literature with special reference to the catalyst systems based on non-noble metals will be provided to assess the likelihood of these nanocatalyst to serve as a commercial grade electrode material for sea water electrolysis.


**Keywords:** metal nitride, nanocatalyst, water electrolysis, hydrogel evolution reaction, oxygen evolution reaction

## 1. Introduction

Low cost, robust alkaline electrolyzers employ non-noble metals as electrodes and allow sustainable electrolysis of water for generation of $H_2$ at a commercial level. The goal is cheaper hydrogen production by minimizing the energy consumption and using low cost cell components. The important factor for consideration here is the use of sea water for $H_2$ generation in the cell environment. Sea-water and ground water are multicomponent natural electrolytes with alkali chloride as the main component. However, the major issue with such water electrolysis is the production of toxic chlorine gas ($Cl_2$) at the anode along with $O_2$ gas. Oxygen evolution reaction (OER) involves a four electron transfer, which makes it much more difficult to catalyze compared to the chlorine evolution reaction (CER), where only a two electron transfer is involved. There have been several attempts to suppress CER during saline water electrolysis such as use of special catalysts that favor OER over CER, coating the anode with protective layers to prevent it from adsorption of chloride ion, and salting out NaCl from the electrolyte solution. Another problem of using sea water or high TDS water is the precipitation of hydroxides of magnesium and calcium on electrode surfaces due to alkaline nature of cathode. Removal of calcium and magnesium prior to electrolysis involve additional cost and generation of solid wastes. Therefore, prevention of precipitation of hydroxides is necessary to run the electrolyzer smoothly for a long period. Recent studies show that formation of $Mg^{2+}$ and $Ca^{2+}$ precipitate on the electrode can be overcome by operating the electrolyzer at near neutral pH, which would also enable the use of low cost earth-abundant electrocatalysts.

## 2. Metal oxide and hydroxide based catalysts for electrolysis

Metal oxides were the first generation of catalysts synthesized and studied to facilitate sustained electrolysis of water for $H_2$ generation. Bennet reported anodes based on $MnO_x$ that exhibit high selectively towards OER over CER in acidic saline water [1]. Since then, many oxides and (oxy)hydroxides (-OOH) of first-row transition metals have been investigated for OER as low-cost alternatives to the noble metals. Kato and coworkers evaluated Mn-based mixed metal oxides quoted on an $IrO_x$/Ti surface and the catalyst system exhibited nearly 100% selectivity towards OER [2]. Strasser and co-workers used NiFe layered double hydroxide nanoplates as OER selective electrocatalysts in seawater. However, the selectivity was limited within over potential range of <480 mV at current density value of 10 $mAcm^{-2}$ [3]. Koper and co-workers used the strategy of depositing a thin $MnO_x$ film onto $IrO_x$ on glassy carbon support that moderately decreased the catalytic activity and strongly shifted the product selectivity from $Cl_2$ towards $O_2$. The $MnO_x$ deposit was catalytically inactive and instead seemed to function as a diffusion barrier that prevented $Cl^-$ ion from reacting on the $IrO_x$ catalyst surface present below, while ensured the transport of water, $H^+$, and $O_2$ between $IrO_x$ and the electrolyte solution required to maintain OER activity [4]. Overall, the issue with oxide and hydroxides was the overpotential value that render the CER as competitive reaction along with OER. As can be seen in the **Table 1** below, the overpotential value with most of the hydroxides were on the higher side.

| Catalysts | Electrolytes | Overpotential (mV) at Specific Current Density | Tafel Slope (mV $dec^{-1}$) |
|---|---|---|---|
| NiOOH | 0.1 M KOH | 375@ 5 $mAcm^{-2}$ | — |
| γ-$Ni_{0.87}Fe_{0.13}OOH$ | 0.1 M KOH | 390@ 10 $mAcm^{-2}$ | — |
| γ-$Ni_{0.75}Fe_{0.25}OOH$ | 0.1 M KOH | 370@ 10 $mAcm^{-2}$ | — |
| NiFe-LDH | 1.0 M KOH | 300@ 10 $mAcm^{-2}$ | 40 |
| NiCo-LDH | 1.0 M KOH | 335@ 10 $mAcm^{-2}$ | 41 |
| FeNi-rGO LDH | 1.0 M KOH | 195@ 10 $mAcm^{-2}$ | 39 |
| FeNi-GO LDH | 1.0 M KOH | 210@ 10 $mAcm^{-2}$ | 40 |
| NiCr-DH | 0.1 M KOH | 310@ 1 $mAcm^{-2}$ | — |
| NiMn-DH | 0.1 M KOH | 380@ 1 $mAcm^{-2}$ | — |
| NiFe-DH | 0.1 M KOH | 270@ 1 $mAcm^{-2}$ | — |
| NiCo-DH | 0.1 M KOH | 500@ 1 $mAcm^{-2}$ | — |
| $Ni(OH)_2$ | 0.1 M KOH | 410@ 1 $mAcm^{-2}$ | — |
| NiCu-DH | 0.1 M KOH | 450@ 1 $mAcm^{-2}$ | — |
| NiZn-DH | 0.1 M KOH | >500@ 1 $mAcm^{-2}$ | — |

**Table 1.**
*The OER overpotential value of metal hydroxide catalysts reported in recent literature. (reproduced with permission from Ref. [5] Copyright 2019 WILEY-VCH Verlag GmbH).*

From the above discussions, it is apparent that major technical challenges that hinder the progress of sea-water/ground-water splitting are the $Cl_2$ gas evolution and the deposition of insoluble $Mg(OH)_2$ and $Ca(OH)_2$ precipitates. Suppression of CER at high current density is possible by using an active electrocatalyst which works for OER below the overpotential of CER. Additionally, use of ultrathin coatings based on $SiO_2$ or $TiO_2$ can selectively block $Cl^-$ ions and thereby suppress the undesirable CER. Precipitation process can be overcome by maintaining the pH at or near neutral value as pKa for the $Mg(OH)_2$ and $Ca(OH)_2$ precipitation reactions are 10.8 and 12.5, respectively. Therefore, it is imperative to develop more advanced electrolyzer systems that can split such waters efficiently and cost-effectively at or near neutral pH environment. Various strategies were adopted to address the issue of CER through development of suitable catalyst system. For example, Dai and coworkers have recently developed a NiFe/$NiS_x$/Ni anode for active, stable and long term seawater electrolysis [6]. The Ni foam was uniformly electrodeposited with NiFe possessing an underneath $NiS_x$ interlayer served as a highly selective OER catalyst for seawater splitting under alkaline condition, the conductive interlayer based on Nickel sulfide provided the stability to the electrode against $Cl^-$ corrosion and degradation. This seawater electrolyzer working under a potential of 2.1 V achieved current density value of 400 $mA/cm^2$ in seawater electrolyte under room temperature conditions. The stability test revealed that no loss in activity was noticeable up to 1000 h. In spite of the significant efforts to develop electrocatalysts for HER and OER, to the best of our knowledge, there is no commercially available electrolyzer that can split sea water or high total dissolved solid (TDS) containing water into $H_2$ and $O_2$. Abundance of Ni is ample superior (90 ppm in nature) compared to other transition metals and the cost is ~4000 times lower compared to that of the benchmark Pt. Therefore, the aim is to design low cost affordable sea water electrolyzer using non-noble metal based nitride/phosphide/sulfide/carbide/graphene nanocatalysts. The chapter especially focuses on the development of nitride based affordable catalyst systems to understand the state of the art and the promise associated with such catalyst system towards catalyzing sustainable sea water or low value water in a sustainable manner.

## 3. Synthetic procedure for metal nitride based catalyst

The most widely used method for the preparation of nanostructured metal nitrides is via heating the corresponding oxides and hydroxides in the presence of different nitrogen sources such as $NH_3$, $N_2$, $NH_2NH_2$, urea, and dicyanamide [7–9]. For the synthesis of binary metal nitrides such as $Mo_2N$, and $Fe_xN$ etc., $NH_3$ is frequently used as the nitrogen source with heat treatment between 400 and 1000°C. The heating rate, gas flow rate and reaction time typically controls the composition of resulting catalyst. For example, heating molybdenum oxide with $NH_3$ at a flow rate of 100 mL/min at 700°C for 2 h produced $Mo_2N$ [10]. In this procedure, the metal salt along with a polymer (PVP) was dispersed in DMF to form an uniform coating on the substrate before heat treatment. As per the SEM and TEM data, the coating of nitride catalyst on the Ni foam was uniform. The procedure was effective to produce crystalline Fe and Ni nitrides. Importantly, optimization in reaction condition is necessary to control the nanostructure of the catalyst, that is important for improved catalytic activity. For example, liquid exfoliation and templating are used to prepare ultrathin 2D nanosheets [11]. Liquid exfoliation is a relatively simpler technique, in which the bulk metal nitride synthesized is added to a high polar solvent such NMP, and the mixture is ultrasonicated to exfoliate the catalyst to nanosheets. For example, a solvent exfoliated atomically thin MoN nanosheet demonstrated improved HER activity compared to that of the as prepared catalyst [12].

Similarly, $N_2$ is another nitrogen source used for synthesis of metal nitride via calcination process and plasma treatment [13]. The ternary metal nitrides are prepared from the ternary metal oxides via treatment with a nitrogen source [14]. In many of the reports, the metal oxides are utilized as the precursor for the corresponding nitrides and the resulting nitride catalysts have mimicked the structure of the oxide precursor. For example, the electrocatalytic Ni-Mo nitride nanotubes were synthesized by heating $NiMoO_4$ nanotubes under $NH_3$ atmosphere at 550°C. The first step involved the synthesis of Ni-Mo bimetallic oxide nanorods under heat treatment in presence of air. Subsequently, the $NiMoO_4$ was converted to NiMoN in presence of $NH_3$. Metal hydroxides are another option for being used as the precursor in metal nitride synthesis. The advantage with the hydroxides is that the metal hydroxides require lower temperature compared to that of the oxide precursors for conversion. For example, $Ni_3FeN$ was successfully synthesized by nitridation reaction with the corresponding double hydroxide precursors in presence of $NH_3$. The $Ni_3Fe$ layered double hydroxides were heated at 400°C to prepare the $Ni_3FeN$ nanoparticles. The SEM images supported a change in surface morphology after ammonia treatment and the TEM displayed lattice spacing of 0.217 nm consistent with the 111 plane of the catalyst supporting the synthesis of $Ni_3FeN$ nanocatalyst.

Some specific procedures include 1-methylimidazole (1-MD)-fixation" strategy to support nano/micro-sized nitrides on carbon materials [15]. Covalent organic frameworks (COFs) were utilized to support $Ni_3N$ nanoparticles through a solid state synthesis to prepare COF-$Ni_3N$ composite [16]. This approach has two advantages, the first one is, this procedure allows nanoscale confinement of the metal nitride catalyst which is otherwise difficult to achieve. The second advantage is a $\pi$-conjugated support further aids the conductivity of nitride catalyst, a property desirable for OER. Though, volatile source rich with nitrogen is treated at high temperature with metal oxides with a programmed temperature ramp is an established procedure to generate the corresponding metal nitrides, plasma treatment is also used in literature to convert $Ni(OH)_2$ to corresponding nitride [17]. This procedure allows the synthesis of catalyst at a relatively low temperature of 250°C using $N_2$-$H_2$ plasma as the source. In the above work, Li et al. utilized the plasma treatment to synthesize $Ni_3N$ nanocatalysts of 30 nm size. The XPS and XRD spectra revealed quantitative conversion of the hydroxides to nitrides in a fairly short duration of 1 h. Nitrates [$Ni(NO_3)_2$] can also be utilized as a source to prepare the nitride catalyst as reported in literature. For example, recently a mixture of $Ni(NO_3)_2{\cdot}6H_2O$ and $(NH_4)_6Mo_7O_{24}{\cdot}4H_2O$ is treated with $NH_3$ as the "N" source to synthesize ammonium nickel molybdate [18]. The XRD data of the resulting catalyst $NiMo_4N_5$ revealed 111 plane accountable to the FCC lattice of Ni, whereas the planes for crystalline phase related to Mo was absent suggesting uniform distribution of the two metals. Derivation of catalyst from a rigid MOF precursor is recently utilized to develop catalyst with controlled nanostructure. For example, Co-$Mo_2N$ was synthesized by heat treatment of ZIF-67/Mo-MOFs-2 at 500°C for 3 h. The resulting catalyst mimicked the shape of MOF precursor [19]. The XPS data showed the peaks assigned to Mo and N present in Mo-N linkage supporting the synthesis of catalyst.

In most of the synthesis, it has been observed that the shape of the precursor material is retained after nitridation in presence of $NH_3$ at high temperature. In some cases, the nitridation also induces porosity in the samples. For example, $Co_3FeN_x$ porous nanowires were synthesized from $Co_3Fe$ double hydroxide nanofibers using $NH_3$ as the nitrogen source at 623 K [20]. The resulting catalysts were effective against both OER and HER and current density values of 20 $mA/cm^2$ and 10 $mA/cm^2$ were achievable at 222 mV and 23 mV overpotentials for OER and HER respectively. Sometimes color can be used as an indicator of the surface coating and functional group conversion. For example, $NiFe(OH)_x$ was grown in-situ on bare Ni foam and the color changed from gray to brown. Further the sample was calcined in presence of $NH_3$ to convert the hydroxides to corresponding nitrides. The color of the surface changed to black suggesting the functional group conversion [21]. In this case, the Ni foam not only serves as a substrate but also acts as a precursor during the hydroxide growth process via redox itching of Fe precursor. A distinct shift in the Ni binding energy from 853 to 855.7 eV was observed in the XPS spectra suggesting conversion of hydroxides (Ni-OH) to the corresponding nitrides (Ni-N). Recently, a non-stoichiometric $NiN_x$ was directly synthesized from Ni foam by plasma treatment. A piece of Ni foam was exposed to $N_2$ plasma initiated by microwave for in-situ growth of nickel nitride nanostructures. The SEM and TEM images displayed change in surface morphology after exposure to plasma. The XPS data displayed a peak at 398 eV accountable to the "N" of Ni-N linkage supporting the formation of nanocatalyst on surface of Ni foam [22].

One of the problems that researchers have frequently faced is the weak bonding between the substrate and nanocatalyst. To address this, a strategy was recently utilized via use of inks. Importantly, the strategy worked with a number of metal catalysts and the method was relatively convenient. In this approach, the metal salts were dissolved in an organic solvent to prepare the ink. Subsequently, the substrate was dipped in the ink to soak the salt on the surface. The sample was then heated at 500°C under $NH_3$ for nitridation. The strategy allowed uniform distribution of the catalyst on surface and the durability of the system was adequate, which will be discussed in the subsequent section. Another challenging aspect of catalyst synthesis is to have a nanocomposite coating of two elements on the metal surface. For example, polymerization-pyrolysis-evaporation strategy was utilized to synthesize nitride doped porous carbon anchored on atomically dispersed $FeN_4$. The synthetic strategy involved two steps; in the first step bimetallic Zn/Fe polyphthalocyanine was synthesized and in the second step the above polymer was pyrolyzed to produce the final catalyst. The HAADF-STEM data revealed uniform distribution of C and N on Fe surface. The catalytic sites were further demonstrated by $^{57}Fe$ Mössbauer transmission spectra [23]. Sputtering technique is also used in literature to deposit metal nitrides on base material such as carbon black and Ni foam. The temperature of the sputtering chamber controlled the stoichiometry of the resulting catalyst. For example, at 90°C $Ni_4N$ was synthesized, whereas at 180°C $Ni_3N$ formed on the surface of carbon cloth at 18 mTorr. Near-edge X-ray absorption fine structure (NEXAFS) spectroscopy revealed the bands for both $Ni_4N$ and $Ni_3N$ supporting the stoichiometric control in the above synthesis. Some other specific modes of synthesis of metal nitrides is by utilizing a reactive ammonia species to decrease the reaction temperature. For example, the manganese and iron oxide were treated with $NaNH_2$ to synthesize the corresponding metal nitrides at 240°C [24].

## 4. Catalytic efficiency of metal nitride based catalysts

Overall, the literature supported that nitrides are more conductive compared to that of the oxides. Owing to the above, metal nitrides exhibit lower overpotential compared to that of the oxides and may be used as bifunctional catalysts for both HER and OER. Doping improves the stability and in some cases the efficiency of the catalyst. Metal alloy can be advantageous as base material. Ni, Fe, Co, Mo are some of the possible low cost metal precursors for utilization as catalyst. Shape of the nanocatalyst also offer possibility of further efficiency improvement. Thin coating of metal nitrides is advantageous compared to the bulk coating. For example, $Ni_3N$ nanosheets exhibited an overpotential value of ~380 mV at 100 $mA/cm^2$ current density, whereas the bulk sample and the corresponding oxide exhibited overpotential value of 600 mV to achieve the above current density. The Tafel slope value for the nanosheet was 45 mV/dec, whereas the bulk catalyst exhibited a Tafel slope value of 85 mV/dec. [25] Recently, a $CuNi_2N$ fabricated on carbon cloth, exhibited 71.4 mV overpotential during HER with a Tafel slope value of 106.5 mV/dec. Long time stability test showed the retention of voltage at 10 $mV/cm^2$ current density for 60 h [26]. A plasma transformed $Ni_3N$ porous nanosheet exhibited 46 mV/dec of Tafel slope value during HER with overpotential marginally higher than that of the Pt/C [27].

Among the metal nitrides studied, ternary metal nitride systems are the most promising, since the coordination number is close to one and the adsorption free energy is close to zero that is most favorable for HER as per the Sabatier's principle [28]. In fact, a large number of ternary metal nitrides have shown high activity towards HER such as NiMo-N [29, 30], CuNi-N [31], CoMo-N [32], $Fe_2Ni_2N$ [33], $ZCo_3FeN$ [34], $NiCo_2N$ [35] and $Ni_3FeN$ [36] Among the non-noble metal systems, Ni-Fe based nitride systems with metallic characteristics, strong absorption of water molecule and unique electronic structure have displayed efficiency towards both HER and OER reaction. Especially, Ni rich compositions have shown promise towards full water splitting. For example, $Ni_3Fe$-N and Ni2Fe2-N have shown efficiency towards HER and OER [37–40]. Morphology of the catalyst and contact with the GC electrode can significantly affect its performance and long term stability. Therefore, in-situ growth is given importance in the later stage of catalyst development. Especially with the Ni based system, Ni foam could serve as one of the most appropriate base electrode since the contact becomes more efficient. Stacking of multiple electrodes can also be utilized to further aid the efficiency. Overall, the literature has shown that effective catalyst systems based on Ni-Fe/Co-N based systems can be formulated and fabricated for overall electrolysis at low overpotential. The Ni based system is able to yield current density value of 100 $mA/cm^2$ at a low overpotential of 100 mV along with superior durability. The already formidable HER activity of Ni based system may be further augured by decorating the catalyst with Pt and further improve the current density to 200 $mA/cm^2$ at 160 mV overpotential [41]. The Co based system are reported to exhibit low activity towards HER since the d band is far from the centre of HER energy level. Though this can be circumvented to some extent by doping with vanadium [42]. Bimetallic systems have invariably demonstrated superior catalytic activities compared to that of their monometallic analog. For example, Ni-Fe [43–45], Ni-Co [46, 47], and Co-Fe nitrides [48] have all exhibited improved catalytic activities compared to that of their monometallic counterparts. $Ni_3FeN$ catalyst materials are one of the leading candidates for use as HER electrocatalysts. Though the mechanism for HER on these nitride surfaces are a matter of intense research, several studies have supported the metallic nature of the catalyst for swift electron transfer necessary for HER. The surface of metal or the "N" that acts as an active centre is still under investigation. OER being a more energy intensive process compared to that of the HER, is more facilitated, when conductive metal nitrides are used instead of oxides. The ratio of metal and "N" influences the electrical conductivity, which subsequently affects the OER efficiency. It was proved in a Co based system ($Co_2N$, $Co_3N$ and $Co_4N$) that, increase in Co amount increases the intrinsic conductivity [49].

Nitrides based on other metals such as Mo and Co have shown activity towards either HER or OER. For example, the binary nitride based on $Mo_2N$ showed adequate HER catalytic activity. A composite of $Mo_2N$-$Mo_2C$ showed enhanced HER activity compared to that of the $Mo_2N$ alone [50]. Similarly, layered conjugation of $MoS_2$ with $MoN_2$ also improved the electrocatalytic activity [51]. Direct growth of $Mo_2N$ on CNT and N doped carbon matrix can be utilized to improve the electrocatalytic activity [52, 53], Recently, $Mo_2N$–$Mo_2C$ heterojunction on the reduced graphene oxide displayed superior HER activity with low onset potentials of 18 mV under basic medium, that is superior to Pt/C electrode in alkaline media at large current densities [54]. In general, it has been observed that Metal-NC system exhibits superior activity compared to that of the metal nitrides only. Just recently, nitride MXene (V-$Ti_4N_3T_x$) based systems have shown superior HER activity [55]. Several specific examples are discussed below to obtain a fair idea about the catalytic activity of such systems.

Recently, Cai and coworkers have synthesized a Ru cluster of ~1 nm size anchored on N doped carbon surface using a one pot procedure [56]. The resulting catalyst displayed activity towards both HER and OER in alkaline (1 M KOH) medium. Importantly, the over-potential values were notably less in case of both the reactions (HER: 15 mV vs. RHE and OER: 285 mV vs. RHE). The durability test showed that the activity remained largely unaffected after 5000 cycles. They hypothesized that the presence of "N" in the catalyst matrix improved the stability and promoted the HER and OER activity. Similarly, Wu and coworkers have immobilized $Co_{5.47}N$ nanoparticles on C-N matrix by utilizing Co based Zeolite framework as the starting material [57]. The frameworks were pyrolyzed at 700°C in presence of $NH_3$ to form the CN nanoparticles in-situ. The catalyst system was effective for both HER and OER and the over-potential values (149 mV for HER and 248 mV for OER) were lower compared to that of the $IrO_2$ benchmark. The catalyst retained ~82% of original current density at an overpotential value of 248 mV after 10 h, which is superior compared to that of the benchmark (56%). The efficiency of nitrides in other cost effective metal systems were analyzed. Catalysts based on $Ni_3N$ nanosheet exhibited adequate OER performance in an alkaline solution and achieved 52.3 mA $cm^{-2}$ current density at relatively low over-potential (350 mV) with small Tafel slope [58]. The nanosheets were prepared by coating an activated carbon cloth with Ni salts and heating the salt coated carbon cloth at 380°C in presence of $NH_3$. The catalyst also exhibited Tafel slope value up to 45 mV/dec.

The Fe based catalyst ($Fe_3N/Fe_4N$) nanoporous film on a conducting Ni-graphene foam displayed low OER overpotential (238 mV) corresponding to current density of 10 $mA/cm^2$ and a low Tafel slope value of 44.5 mV/dec along with high 96.7% faradaic yield. These numbers were superior compared to that of the benchmark $IrO_2$ and attributed to high electron transfer and surface area of the catalyst [59]. The OER overpotential of the nitrides were far superior than the corresponding oxide as shown below. In fact, a similar trend was noticed with Co based nitride ($Co_4N$) catalyst system. In which a current density value of 10 mA/$cm^2$ was achieved at an overpotential of 257 mV with small Tafel slope [60]. Most of the reports on nitride based system revealed the bi-functional nature of the catalyst. These catalysts were effective against both HER and OER under basic conditions. Additionally, most of the catalyst based on the nitrides displayed lower overpotential value compared to that of the oxide based systems. The lower overpotential values of these systems were assigned to their higher conductivity values that resulted from the metallic character arising out of the overlap between Ni-3d and N-2p orbitals in the catalysts. The second promising aspect was their bifunctional behavior, that resulted from the tendency of "N" to donate electron more easily compared to that of the "O" and polarization associated with the shift of "H" from "O" to "N", a key step during HER [61]. The adsorption energy value of "H" on nitride based catalyst system was much lower compared to that of the $H_2O$ and similar to that of the Pt-C bench mark that facilitated the rate of HER. The nitride systems based on the metal alloys exhibited lower overpotential values compared to that of their single metal counterparts. For example, the overpotential value of CoFe(3:1)-N was ~150 mV lower compared to that of the Co-N or Fe-N for OER. The Tafel slope value was approximately 3 times lower in case of alloy nitrides further supporting the efficiency of these systems towards OER. The ratio between the two metals in the alloy played an important role, while determining the catalytic efficiency [62]. The stability of these catalysts were also excellent as only 4.5% decrease in efficiency was noticed after 10 h of OER. The spatial arrangement of these alloy nitride nanostructures on the base electrode also displayed a variation in catalytic efficiency. For example, NiFe-N grown on Z plane from the electrode surface displayed 277 mV overpotential at 100 $mV/cm^2$ current density and 337 mV at 500 $mV/cm^2$, which was lower compared to that of the $IrO_2$ benchmark (542 mV at 500 $mV/cm^2$) [63]. The Tafel slope value for the above catalyst system was 58.6 mV/dec.

Further studies have shown that, the shape of the nanocatalyst controls the efficiency to a substantial extent. Report based on Rh nanocrystals have shown that, the benzoid structures exhibited lower overpotential in both OER and HER compared to that of the tetrahedral structures [64]. However, the deviation was within 100 mV in case of OER. Interestingly, the Tafel slope value displayed a strong improvement on optimization of the shape. The value for the benzoid structure was 87 mV/dec, whereas the value for the tetrahedral structure was 205 mV/dec. Other factors such as doping of metal nanoparticles on the catalyst can also be used as a procedure to further improve the overpotential value and dependency of current on the overpotential. Recently, a cobalt nitride based nanofiber system was doped with Ir nanoparticle. The resulting catalyst exhibited much lower Tafel slope value compared to that of the undoped system. The overpotential value of the doped system was also lower compared to that of the base nitride system [65]. Similarly, a chromium doped Co-N system exhibited a Tafel slope value of 38.1 mV/dec and retained current density up to 200 h [66]. The system displayed an overpotential value of 99 mV at 100 $mV/cm^2$. The current versus potential curve was superimposable

| Catalyst Composition | Electrolyte Used | Overpotential (mV) @ 10 mA $cm^{-2}$ Current Density | Tafel Slope (mV $dec^{-1}$) | Reference |
|---|---|---|---|---|
| $Ni_3N_{1-x}$/NF | 1.0 M KOH | 55 | 54 | [68] |
| $Ni_3FeN$/NF | | 75 | 98 | [69] |
| $Ni_3FeN$ | | 158 | 42 | [70] |
| NiMoN/CC | | 109 | 95 | [71] |
| MoON/CC | | 146 | 101 | [55] |
| $Ni_3N$/CC | | 208 | 113 | [55] |
| $Mo_2N$ | | 353 | 108 | [72] |
| $Ni_3N$ | | 96 | 120 | [73] |
| $Ni_3N$/NF | | 121 | 109 | [74] |
| NiCoN nanowires | | 145 | 105.2 | [75] |
| CoN nanowires | | 97 | 93.9 | [76] |
| $Co_{5.47}N$@N-C | | 149 | 86 | [77] |
| $Ni_3N$ nanorods | | 305 | 197.5 | [78] |
| Co-$Ni_3N$ nanorods | | 194 | 156 | [62] |
| $Ni_3N$ nanosphere | | 185 | — | [79] |

**Table 2.**
*The table summarizes the values of the overpotential and Tafel slope values for different catalyst compositions. Ref. [67] (supporting information).*

after 1000 cycles. The values were superior compared that of the Pt benchmark. A common study involving various earth abundant metal catalysts revealed that Tafel slope value of the non-stoichiometric nickel nitride catalyst coated on Ni ($Ni_3N_x$) foam is the lowest among various catalyst systems studied (**Table 2**).

## 5. Durability of metal nitride based catalysts

Published literature provides a fair idea about the durability of nitride based catalysts under electrolysis conditions [80]. We are revisiting several compositions and their lifetime in the following section. The NiMoN nanotube displayed OER efficiency equivalent to that of the $IrO_2$ reference, as the current versus potential curve almost superimposed in both the cases. The catalyst displayed an overpotential of 295 mV at 10 mA/cm$^2$ current density. The stability was checked by performing 1000 cycles of CV scanning within 1.036 to 1.636 V at a scan rate of 50 mV/sec. The CV traces remained unchanged after 1000 cycles supporting the stability of nanocatalyst. The stability curve at 295 mV overpotential also retained the current density up to 20 h [30]. The stability comparison between monometallic and bimetallic metal nitrides revealed that later is more stable under HER condition. For example, the CoN under constant overpotential of 100 mV retained current density up to 24 h. The $NiCo_2N$ exhibited current retention up to 48 h under HER condition at lower overpotential (50 mV). The overpotential values of the bimetallic system was also superior compared to that of the single metal system. The electrolysis was carried out under basic conditions (1 M KOH). These enhanced activity is attributed to the synergistic effect of both the metals that increased the conductivity and facilitated the charge transfer necessary for efficient HER [81]. The iron nickel alloy nitride systems have shown most comfortable overpotentials towards OER as described in the previous section. The stability of these class of catalysts becomes important as the plan is to pursue similar composition for our approach. Successive OER and HER was carried out with the same catalyst for 30 h each at an overpotential of 100 and − 100 mV respectively. The sample exhibited adequate stability and the crystallinity also remained intact after the reactions as displayed from the XRD.

The durability test also revealed that the current was maintained up to 400 h under basic conditions at 10 mA/cm$^2$ [21]. A Ru nitride (non-stoichiometric) coated on carbon black catalyst retained current density (10 mA/cm2) up to 50 h, even though the electrolyte was changed two times in between the period. At the same time, the reference could sustain the current only for 10 h and the current density started decreasing to 1 mA/cm$^2$ after 2 h. Importantly, the chronoamperometry trace was identical to the original one after 50 h of catalysis supporting stability. Similarly, the overpotential value remained intact after 1000 cycles of LSV, whereas the catalyst without nitride coating degraded with a much higher overpotential [82]. An electrolysis study with $Co_2N$, $Co_3N$ and $Co_4N$ showed that all three catalyst systems exhibited current retention up to 3.3 h at 437 mV overpotential in 0.1 M KOH. The oxygen production value was similar to that of the theoretically calculated value for 1 h. The stability of Co based system was lower compared to that of the Ni/Fe based system. The chronoamperometry trace after 1000 cycle matched with that of the initial one. Overall, Ni in conjugation with other transition metals in the nitride form have exhibited adequate stability with a low overpotential [83]. The overpotential of $Ni_3Fe$ was marginally lower than that of the $Ni_3Co$ nitride and much lower than that of the $Ni_3Mn$ nitride. Both $Ni_3FeN$ and $Ni_3CoN$ exhibited retention of potential (1.55 V) under 100 A/g current density. However, the $Ni_3MnN$ could sustain the activity up to 16 h under OER conditions. The $Ni_3FeN$ exhibited stability under HER conditions with efficiency. The LSV curves were

| Catalyst | Morphology | Overpotential (mV) | Onset Potential (V vs. RHE) | Tafel Slope (mV dec$^{-1}$) | Stability |
|---|---|---|---|---|---|
| $Ni_3N$ | 2D-sheets | 250 | 1.55 | 45 | Excellent |
| $Ni_3N$/CC | 3D-sheets | 190 | 1.36 | 112 | Good |
| $Ni_3N$/NF | 3D-sheets | 50 | 1.39 | 60 | Excellent |
| $Fe_2Ni_2N$/CNT | 1D-sheets | 282 | 1.47 | 38 | Good |
| $Ni_3FeN$ | 2D-sheets | 300 | 1.35 | 51 | Excellent |
| $Ni_3FeN$/NW | 3D-sheets | 200 | 1.34 | 40 | Excellent |
| $Ni_3FeN$/CC | 3D-sheets | 240 | 1.45 | 59 | Excellent |
| $Ni_3FeN$/CC | 3D-sheets | 105 | 1.33 | 72 | Excellent |
| $Ni_3FeN$/rGO | 3D-sheets | 280 | 1.42 | 90 | Good |
| NiCoN/NW | 1D-nanowires | 360 | 1.53 | 45 | Good |
| $Ni_3CoN$ | 2D-sheets | 340 | 1.52 | 55 | Excellent |
| $NiCo_2N$/NF | 3D-sheets | 180 | 1.35 | 69 | Good |
| $Ni_3MnN$ | 2D-sheets | 320 | 1.60 | 64 | Excellent |
| NiMoN/CF | 1D-nanofibres | 210 | 1.32 | 55 | Good |
| NiMoN/NF | 3D-foams | 218 | 1.35 | 55 | Excellent |

*The electrolytes used in above cases were 1.0 M KOH or NaOH.*

**Table 3.**
*Ni based nitride catalyst efficiency and stability under OER conditions (reproduced with permission from ref. [5] copyright 2019 WILEY-VCH Verlag GmbH).*

repeatable after 1000 cycles [84]. Finally, a summary of the Ni based nitride catalyst efficiency under OER condition is presented to obtain a generic understanding about the capability of these class of materials. The overpotential value is distributed within 50 to 360 mV for OER under basic conditions with stability is termed as good to excellent suggesting the chances of developing a Ni nitride based catalyst for commercial grade electrolysis is fairly high (**Table 3**).

## 6. Effect of salt on electrolysis efficiency

Though most of the articles have studied pure water electrolysis using metal nitrides as the catalyst, a few reports suggests that metal nitrides are promising as the catalyst for sea water electrolysis due to their corrosion resistant and electrically conductive properties [85]. The overpotential values marginally increased, when NaCl was added to the water or seawater was directly used as the electrolyte. In case of NiMoN, the overpotential values increased from 130 to 160 mV, when seawater replaced the pure water as the electrolyte. Interestingly, addition of NaCl to the solution, didn't change the HER overpotential value significantly at 500 mA/cm$^2$ current density. Similarly, the OER overpotential value increased from 340 to 355 mV at 500 mA/cm$^2$ current density in presence of salt and the value further increased to 365 mV in presence of sea water suggesting, though the salt water affects the overpotential, the increase is not that significant. The durability tests were also conducted in presence of NaCl and sea water. The overpotential value exhibited minor increase after 100 h at 500 mA/cm$^2$ current density suggesting the catalyst may be a viable option for exploring possibility as a commercial catalyst towards further research and development [63]. Recently, a $Mo_5N_6$ nanosheet along

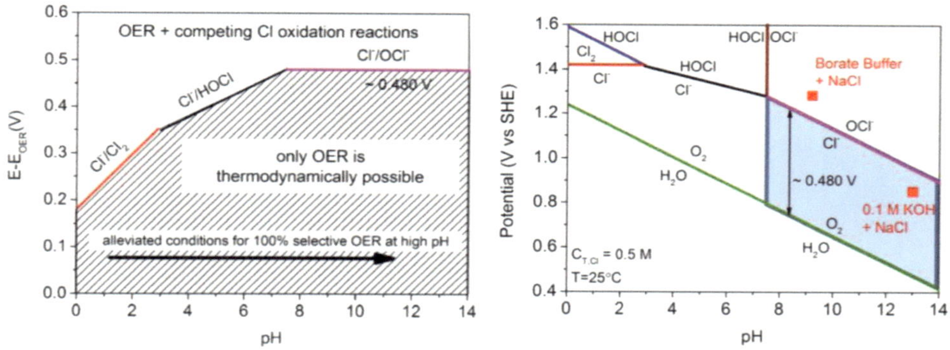


**Figure 1.**
*The overpotential versus pH for the selective OER and the form of $Cl^-$ conversion in presence different pH conditions (reproduced with permission from ref. [3] copyright 2016 WILEY-VCH Verlag GmbH).*

with several metal nitrides were studied for their efficiency towards sea water electrolysis. As per the report, the HER overpotential was least affected by the presence of sea water. Similarly, the current retention was also 100% after a 100 h cycle at an applied potential of 310 mV [86]. Similarly, a NiNS based bifunctional catalyst system was also studied for sea water electrolysis. A current density value of 48.3 $mAcm^{-2}$ at 1.8 V was achieved for overall sea water electrolysis. The current density value marginally decreased from 15 to 13 $mA/cm^2$ over 12 h period of electrolysis. To conclude, the overpotential values associated with these metal nitrides based catalyst systems are adequate to carry out electrolysis without affecting $Cl^-$ ion. In case of basic pH electrolysis, the allowed overpotential is 450 mV, whereas under acidic conditions, the overpotential is limited to ~250 mV. Moreover, the $Cl^-$ is expected to release as $Cl_2$ gas under acidic conditions, whereas under basic conditions, the hypochlorate ion is going to precipitate as salt and may not affect the efficiency to a certain extent (**Figure 1**). Therefore, considering the overall scenario, basic electrolysis system may be one of the safer option when sea water is a part of the electrolyte.

## 7. Conclusions

In conclusion, the chapter enlightens the necessities of nitride based catalysts for sea and ground water electrolysis in the preliminary section. Subsequently, the synthetic strategy utilized for these metal nitride nanocatalysts, their efficiency and efficacy in presence of sea water as electrolyte is summarized. To ensure $H_2$ energy becomes one of the commercially viable energy source for consumption, a sustainable generation mode is highly desirable. Metal nitride based catalysts have shown promise to fulfill the same with numerous research publications providing scientific data in support of the above. Though, still commercial implementation is yet to be achieved with these class of catalyst materials, we can safely assume that in near future a realistic design of electrolyzers possessing these nitride nanocatalyst modified electrodes may be available for commercial production.

## Author details

Akhoury Sudhir Kumar Sinha* and Umaprasana Ojha
Rajiv Gandhi Institute of Petroleum Technology, Jais,
Amethi, Uttar Pradesh-229304, India

*Address all correspondence to: asksinha@rgipt.ac.in

## References

[1] Bennett J E, Electrodes for generation of hydrogen and oxygen from seawater Int. J. Hydrogen Energy 1980;5:401-408. DOI: 10.1016/0360-3199(80)90021-X

[2] Kato Z, Bhattarai J, Kumagai N, Izumiya K, Hashimoto K. Durability enhancement and degradation of oxygen evolution anodes in seawater electrolysis for hydrogen production. Appl. Surface Sci 2011;257: 8230-8236. DOI: 10.1016/j.apsusc.2010.12.042

[3] Dionigi F, Reier T, Pawolek Z, Gliech M, Strasser P. Design Criteria, Operating Conditions, and Nickel–Iron Hydroxide Catalyst Materials for Selective Seawater Electrolysis ChemSusChem. 2016;9:962-972. DOI: 10.1002/cssc.201501581

[4] Vos JG, Wezendonk TA, Jeremiasse AW, Koper MTM. MnOx/IrOx as Selective Oxygen Evolution Electrocatalyst in Acidic Chloride Solution J. Am. Chem. Soc. 2018;140:10270-10281. DOI: 10.1021/jacs.8b05382

[5] Tareen AK, Priyanga GS, Khan K, Pervaiz E, Thomas T, Yang M, Nickel-Based Transition Metal Nitride Electrocatalysts for the Oxygen Evolution Reaction. ChemSusChem. 2019;12: 3941-3954. DOI: 10.1002/cssc.201900553

[6] Kuang Y, Kenney MJ, Meng Y, Hung WH, Liu Y, Huang JE, Prasanna R, Li P, Li Y, Wang L, Lin MC, McGehee MD, Sun X, Dai H. Solar-driven, highly sustained splitting of seawater into hydrogen and oxygen fuels. Proc. National Acad. Sci. 2019;116: 6624-6629. DOI: 10.1073/pnas.1900556116

[7] Zhang Y, Ouyang B, Xu J, Jia G, Chen S, Rawat R, Fan-Jin H. Rapid Synthesis of Cobalt Nitride Nanowires: Highly Efficient and Low-Cost Catalysts for Oxygen Evolution. Angew. Chem. Int. Ed. 2016;55:8670. DOI: 10.1002/anie.201604372

[8] Ma L, Ting RL, Molinari V, Giordano C, Yeo B. Efficient hydrogen evolution reaction catalyzed by molybdenum carbide and molybdenum nitride nanocatalysts synthesized via the urea glass route. J. Mater. Chem. A. 2015;3:8361-8368. DOI: 10.1039/C5TA00139K

[9] Shalom M, Molinari V, Esposito D, Clavel G, Ressnig D, Giordano C, Antonietti M. Sponge-like nickel and nickel nitride structures for catalytic applications. Adv. Mater. 2014;26:1272-1276. DOI: 10.1002/adma.201304288

[10] Yu F, Zhou H, Zhu Z, Sun J, He R, Bao J, Chen S, Ren Z. Three-Dimensional Nanoporous Iron Nitride Film as an Efficient Electrocatalyst for Water Oxidation. ACS Catal. 2017;7:2052-2057. DOI: 10.1021/acscatal.6b03132

[11] Xiong J, Cai W, Shi W, Zhang X, Li J, Yang Z, Feng L, Cheng H. Salt-templated synthesis of defect-rich MoN nanosheets for boosted hydrogen evolution reaction. J. Mater. Chem. A. 2017;5: 24193-24198. DOI: 10.1039/C7TA07566A

[12] Xie J, Li S, Zhang X, Zhang J, Wang R, Zhang H, Yi Pan B, Xie Y. Atomically-thin molybdenum nitride nanosheets with exposed active surface sites for efficient hydrogen evolution. Chem. Sci. 2014;5:4615-4620. DOI: 10.1039/C4SC02019G

[13] Reger N-C, Balla V-K, Das M, Bhargava A-K. Wear and corrosion properties of in-situ grown zirconium nitride layers for implant applications. Surf. Coat. Technol. 2018;334:357-364. DOI: 10.1016/j.surfcoat.2017.11.064

[14] Bhattacharyya S, Kurian S, Shivaprasad S-M, Gajbhiye N-S. Synthesis and magnetic characterization of CoMoN2 nanoparticles. J. Nanopart. Res. 2010;12:1107-1116. DOI: 10.1007/s11051-009-9639-5

[15] Meng M, Yan H, Jiao Y, Wu A, Zhang X, Wanga R, Tian C. A “1-methylimidazole-fixation” route to anchor small-sized nitrides on carbon supports as non-Pt catalysts for the hydrogen evolution reaction. RSC Adv. 2016;6:29303-29307. DOI: 10.1039/C5RA27490G

[16] Nandi S, Singh S-K, Mullangi D, Illathvalappil R, George L, Vinod C-P, Kurungot S, Vaidhyanathan R. Low Band Gap Benzimidazole COF Supported Ni3N as Highly Active OER Catalyst. Adv. Energy Mater. 2016;6:1601189. DOI:10.1002/aenm.201601189

[17] Li G, Wu X, Guo H, Guo Y, Chen H, Wu Y, Zheng J, Li X. Plasma Transforming $Ni(OH)_2$ Nanosheets into Porous Nickel Nitride Sheets for Alkaline Hydrogen Evolution. ACS Appl. Mater. Interfaces 2020;12:5951-5957. DOI: 10.1021/acsami.9b20887

[18] Wang T, Wang X, Liu Y, Zheng J, Li X. A Highly Efficient and Stable Biphasic Nanocrystalline Ni-Mo-N Catalyst for Hydrogen Evolution in Both Acidic and Alkaline Electrolytes. Nano Energy 2016;22:111-119. DOI: 10.1016/j.nanoen.2016.02.023

[19] Shi X, Wu A, Yan H, Zhang L, Tian C, Wang L, Fu H. A “MOFs plus MOFs” strategy toward Co–Mo2N tubes for efficient electrocatalytic overall water splitting. J. Mater. Chem. A 2018;6: 20100-20109. DOI: 10.1039/C8TA07906D

[20] Wang Y, Liu D, Liu Z, Xie C, Huo J, Wang S. Porous Cobalt-Iron Nitride Nanowires as Excellent Bifunctional Electrocatalysts for Overall Water Splitting. Chem. Commun. 2016;52:12614-12617. DOI: 10.1039/C6CC06608A

[21] Zhang B, Xiao C, Xie S, Liang J, Chen X, Tang Y. Iron-Nickel Nitride Nanostructures in Situ Grown on Surface-Redox-Etching Nickel Foam: Efficient and Ultrasustainable Electrocatalysts for Overall Water Splitting. Chem. Mater. 2016;28: 6934-6941. DOI: 10.1021/acs.chemmater.6b02610

[22] Liu B, He B, Peng HQ, Zhao Y, Cheng J, Xia J, Shen J, Ng TW, Meng X, Lee CS, Zhang W. Unconventional Nickel Nitride Enriched with Nitrogen Vacancies as a HighEfficiency Electrocatalyst for Hydrogen Evolution. Adv. Sci. 2018;5: 1800406. DOI: 10.1002/advs.201800406

[23] Pan Y, Liu S, Sun K, Chen X, Wang B, Wu K, Cao X, Cheong WC, Shen R, Han A, Chen Z, Zheng L, Luo J, Lin Y, Liu Y, Wang D, Peng Q, Zhang Q, Chen C, Li Y, Bimetallic Zn/Fe Polyphthalocyanine Derived Single-Atom Fe-N4 Catalytic Site:A Superior Tri-Functional Catalyst for Overall Water Splitting and Zn-Air Battery. Angew. Chem. Int. Ed. 2018, 57, 8614-8618. DOI: 10.1002/anie.201804349

[24] Miura, A, Takei T, Kumada N. Low-Temperature Nitridation of Manganese and Iron Oxides Using NaNH2 Molten Salt. Inorg. Chem. 2013, 52, 20, 11787-11791. DOI: 10.1021/ic401951u

[25] Xu K, Chen P, Li X, Tong Y, Ding H, Wu X, Chu W, Peng Z, Wu C, Xie Y. Metallic Nickel Nitride Nanosheets Realizing Enhanced Electrochemical Water Oxidation. J. Am. Chem. Soc. 2015;137:4119-4125. DOI: 10.1021/ja5119495

[26] Xu Z, Liu Z, Li N, Tang G, Zheng G, Zhu C, Chen Y, Wang L, Huang Y, Li L, Zhou N, Hong J, Chen Q, Zhou H. A Thermodynamically Favored Crystal

Orientation in Mixed Formamidinium/Methylammonium Perovskite for Efficient Solar Cells. Adv. Mater. 2019;31:1900390. DOI: 10.1002/adma.201900390

[27] Li G, Wu X, Guo H, Guo Y, Chen H, Wu Y, Zheng J, Li X. Plasma Transforming Ni(OH)2 Nanosheets into Porous Nickel Nitride Sheets for Alkaline Hydrogen Evolution. ACS Appl. Mater. Interfaces. 2020;12:5951-5957. DOI: 10.1021/acsami.9b20887

[28] Gao G, Bottlea S, Du A. Understanding the activity and selectivity of single atom catalysts for hydrogen and oxygen evolution via ab initial study. Catal. Sci. Technol. 2018;8:996-1001. DOI: 10.1039/C7CY02463K

[29] Chen W-F, Sasaki K, Ma C, Frenkel A-I, Marinkovic N, Muckerman J-T, Zhu Y, Adzic R-R. Hydrogen-Evolution Catalysts Based on Non-Noble Metal Nickel–Molybdenum Nitride Nanosheets. Angew. Chem. Int. Ed. 2012;51:6131-6135. DOI:10.1002/anie.201200699

[30] Yin Z, Sun Y, Zhu C, Li C, Zhang X, Chen Y. Bimetallic Ni–Mo nitride nanotubes as highly active and stable bifunctional electrocatalysts for full water splitting. J. Mater. Chem. A. 2017;5:13648-13658. DOI: 10.1039/C7TA02876H

[31] Ma Y, He Z, Wu Z, Zhang B, Zhang Y, Ding S, Xiao C. Galvanic-replacement mediated synthesis of copper–nickel nitrides as electrocatalyst for hydrogen evolution reaction. J. Mater. Chem. A. 2017;5:24850-24858. DOI: 10.1039/C7TA08392K

[32] Cao B, Veith G-M, Neuefeind J-C, Adzic R-R, Khalifah P-G. Mixed Close-Packed Cobalt Molybdenum Nitrides as Non-noble Metal Electrocatalysts for the Hydrogen Evolution Reaction. J. Am. Chem. Soc. 2013;135:19186-19192. DOI: 10.1021/ja4081056

[33] Jiang M, Li Y, Lu Z, Sun X, Duan X. Binary nickel–iron nitride nanoarrays as bifunctional electrocatalysts for overall water splitting. Inorg. Chem. Front.2016;3:630-634. DOI: 10.1039/C5QI00232J

[34] Wang Y, Liu D, Liu Z, Xie C, Huo J, Wang S. Porous cobalt–iron nitride nanowires as excellent bifunctional electrocatalysts for overall water splitting. Chem. Commun. 2016;52:12614-12617. DOI: 10.1039/C6CC06608A

[35] Wang Y, Zhang B, Pan W, Ma H, Zhang J. 3D Porous Nickel–Cobalt Nitrides Supported on Nickel Foam as Efficient Electrocatalysts for Overall Water Splitting. ChemSusChem. 2017;10:4170. DOI: 10.1002/cssc.201701456

[36] Jia X, Zhao Y, Chen Y, Shang L, Shi R, Kang X, Waterhouse G, Wu L, Tung C, Zhang T. Ni3FeN Nanoparticles Derived from Ultrathin NiFe-Layered Double Hydroxide Nanosheets: An Efficient Overall Water Splitting Electrocatalyst. Adv. Energy Mater. 2016;6:1502585. DOI: 10.1002/aenm.201502585

[37] Liu Z, Tan H, Xin J, Duan J, Su X, Hao P, Xie J, Zhan J, Zhang J, Wang J, Liu H. Metallic Intermediate Phase Inducing Morphological Transformation in Thermal Nitridation: Ni3FeN-Based Three-Dimensional Hierarchical Electrocatalyst for Water Splitting. ACS Appl. Mater. 2016;10:3699-3706. DOI: 10.1021/acsami.7b18671

[38] Zhang B, Xiao C, Xie S, Liang J, Chen X, Tang Y. Iron–Nickel Nitride Nanostructures in Situ Grown on Surface-Redox-Etching Nickel Foam: Efficient and Ultrasustainable Electrocatalysts for Overall Water Splitting. Chem. Mater. 2016;28:6934-6941. DOI: 10.1021/acs.chemmater.6b02610

[39] Gu Y, Chen S, Ren J, Jia Y, Chen C, Komarneni S, Yang D, Yao X. Electronic Structure Tuning in Ni3FeN/r-GO Aerogel toward Bifunctional Electrocatalyst for Overall Water Splitting. ACS Nano. 2018;12:245-253. DOI: 10.1021/acsnano.7b05971

[40] Wang Y, Xie C, Liu D, Huang X, Huo J, Wang S. Nanoparticle-Stacked Porous Nickel–Iron Nitride Nanosheet: A Highly Efficient Bifunctional Electrocatalyst for Overall Water Splitting. ACS Appl. Mater. Interfaces. 2016;8:18652-18657. DOI: 10.1021/acsami.6b05811

[41] Wang Y, Chen L, Yu X, Wang Y, Zheng G. Superb Alkaline Hydrogen Evolution and Simultaneous Electricity Generation by Pt-Decorated Ni3N Nanosheets. Adv. Energy Mater. 2017;7:1601390. DOI: 10.1002/aenm.201601390

[42] Chen Z, Song Y, Cai J, Zheng X, Han D, Wu Y, Zang Y, Niu S, Liu Y, Zhu J, Liu X, Wang G. Tailoring the d-Band Centers Enables Co 4 N Nanosheets To Be Highly Active for Hydrogen Evolution Catalysis. Angew. Chem. Int. Ed. 2018;57:5076-5080

[43] Gu Y, Chen S, Ren J, Jia Y, Chen C, Komarneni S, Yang D, Yao X. Electronic Structure Tuning in Ni3FeN/r-GO Aerogel toward Bifunctional Electrocatalyst for Overall Water Splitting. ACS Nano. 2018;12:245-253. DOI:10.1021/acsnano.7b05971

[44] Wang Y, Xie C, Liu D, Huang X, Huo J, Wang S. Nanoparticle-Stacked Porous Nickel–Iron Nitride Nanosheet: A Highly Efficient Bifunctional Electrocatalyst for Overall Water Splitting. ACS Appl. Mater. Interfaces. 2016;8:18652-18657. DOI:10.1021/acsami.6b05811

[45] Chen Q, Wang R, Yu M, Zeng Y, Lu F, Kuang X, Lu X, Acta E. Bifunctional Iron–Nickel Nitride Nanoparticles as Flexible and Robust Electrode for Overall Water Splitting. 2017;247:666-673. DOI:10.1016/j.electacta.2017.07.025

[46] Li S, Wang Y, Peng S, Zhang L, Al-Enizi A-M, Zhang H, Sun X, Zheng G. Co–Ni-Based Nanotubes/Nanosheets as Efficient Water Splitting Electrocatalysts. Adv. Energy Mater. 2016;6:1501661. DOI: 10.1002/aenm.201501661

[47] Ray C, Lee S-C, Jin B, Kundu A, Park J-H, Jun S-C. Conceptual design of three-dimensional CoN/Ni3N-coupled nanograsses integrated on N-doped carbon to serve as efficient and robust water splitting electrocatalysts. J. Mater. Chem. A. 2018;6:4466-4476. DOI: 10.1039/C7TA10933D

[48] Wang Y, Liu D, Liu Z, Xie C, Huoa J, Wang S. Porous cobalt–iron nitride nanowires as excellent bifunctional electrocatalysts for overall water splitting. Chem. Commun. 2016;52:12614-12617. DOI: 10.1039/C6CC06608A

[49] Chen P, Xu K, Tong Y, Li X, Tao S, Fang Z, Chu W, Wu X, Wu C. Cobalt nitrides as a class of metallic electrocatalysts for the oxygen evolution reaction. Inorg. Chem. Front. 2016;3:236-242. DOI: 10.1039/C5QI00197H

[50] Chen W-F, Iyer S, Iyer S, Sasaki K, Wang C-H, Zhu Y, Muckerman J-T, Fujita E. Biomass-derived electrocatalytic composites for hydrogen evolution. Energy Environ. Sci. 2013;6:1818-1826. DOI: 10.1039/C3EE40596F

[51] Ojha K, Saha S, Banerjee S, Ganguli A-K. Efficient Electrocatalytic Hydrogen Evolution from MoS2-Functionalized Mo2N Nanostructures. ACS Appl. Mater. Inter. 2017;9:19455-19461. DOI: 10.1021/acsami.6b10717

[52] Youn D-H, Han S, Kim J-Y, Kim J-Y, Park H, Choi S-H, Lee S-J. Highly Active and Stable Hydrogen Evolution Electrocatalysts Based on Molybdenum Compounds on Carbon Nanotube–Graphene Hybrid Support. ACS Nano. 2014;8:5164-5173. DOI: 10.1021/nn5012144

[53] Lv Z, Tahir M, Lang X, Yuan G, Pan L, Zhang X, Zou J. Well-dispersed molybdenum nitrides on a nitrogen-doped carbon matrix for highly efficient hydrogen evolution in alkaline media. J. Mater. Chem. A. 2017;5:20932-20937. DOI: 10.1039/C7TA06981B

[54] Yan H, Xie Y, Jiao Y, Wu A, Tian C, Zhang X, Wang L, Fu H. Holey Reduced Graphene Oxide Coupled with an Mo2N–Mo2C Heterojunction for Efficient Hydrogen Evolution. Adv. Mater. 2018;30:1704156. DOI: 10.1002/adma.201704156

[55] Djire A, Wang X, Xiao C, Nwamba O-C, Mirkin M-V, Neale N-R. Basal Plane Hydrogen Evolution Activity from Mixed Metal Nitride MXenes Measured by Scanning Electrochemical Microscopy. Adv. Funct. Mater. 2020:2001136. DOI: 10.1002/adfm.202001136

[56] Hu H, Kazim F-M-D, Zhang Q, Qu K, Yang Z, Cai W. Nitrogen Atoms as Stabilizers and Promoters for Ru-Cluster-Catalyzed Alkaline Water Splitting. Chem. Cat. Chem. 2019;11:4327. DOI: 10.1002/cctc.201900987

[57] Chen Z, Ha Y, Liu Y, Wang H, Yang H, Xu H, Li Y, Wu R. In Situ Formation of Cobalt Nitrides/Graphitic Carbon Composites as Efficient Bifunctional Electrocatalysts for Overall Water Splitting. ACS Appl. Mater. Interfaces. 2018;10:7134-7144. DOI: 10.1021/acsami.7b18858

[58] Xu K, Chen P, Li X, Tong Y, Ding H, Wu X, Chu W, Peng Z, Wu C, Xie Y. Metallic Nickel Nitride Nanosheets Realizing Enhanced Electrochemical Water Oxidation. J. Am. Chem. Soc. 2015;137:4119-4125. DOI: 10.1021/ja5119495

[59] Yu F, Zhou H, Zhu Z, Sun J, He R, Bao J, Chen S, Ren Z. Three-Dimensional Nanoporous Iron Nitride Film as an Efficient Electrocatalyst for Water Oxidation. ACS Catal. 2017;7:2052-2057. DOI: 10.1021/acscatal.6b03132

[60] Chen P, Xu K, Fang Z, Tong Y, Wu J, Lu X, Peng X, Ding H, Wu C, Xie Y. Metallic Co4N Porous Nanowire Arrays Activated by Surface Oxidation as Electrocatalysts for the Oxygen Evolution Reaction. Angew. Chem. Int. Ed. 2015; 54:14710-14714. DOI: 10.1002/anie.201506480

[61] Cao L, Luo Q, Liu W, Lin Y, Liu X, Cao Y, Zhang W, Wu Y, Yang J, Yao T, Wei S. Identification of single-atom active sites in carbon-based cobalt catalysts during electrocatalytic hydrogen evolution. Nat. Catal. 2019;2:134-141. DOI: 10.1038/s41929-018-0203-5

[62] Liu T, Li M, Bo X, Zhou M. Comparison Study toward the Influence of the Second Metals Doping on the Oxygen Evolution Activity of Cobalt Nitrides. ACS Sustainable Chem. Eng. 2018;6:11457-11465. DOI: 10.1021/acssuschemeng.8b01510

[63] Yu L, Zhu Q, Song S, McElhenny B, Wang D, Wu C, Qin Z, Bao J, Yu Y, Chen S, Ren Z. Non-noble metal-nitride based electrocatalysts for high-performance alkaline seawater electrolysis. Nat Commun. 2019;10:5106. DOI: 10.1038/s41467-019-13092-7

[64] Zhang N, Shao Q, Pi Y, Guo J, Huang X. Solvent-Mediated Shape Tuning of Well-Defined Rhodium Nanocrystals for Efficient Electrochemical Water Splitting.

Chem. Mater. 2017;29:5009-5015. DOI: 10.1021/acs.chemmater.7b01588

[65] Cho S-H, Yoon K-R, Shin K, Jung J-W, Kim C, Cheong J-Y, Youn D-Y, Song S-W, Henkelman G, Kim I-D. Synergistic Coupling of Metallic Cobalt Nitride Nanofibers and IrOx Nanoparticle Catalysts for Stable Oxygen Evolution. Chem. Mater. 2018;30:5941-5950. DOI: 10.1021/acs.chemmater.8b02061

[66] Yao N, Li P, Zhou Z, Zhao Y, Cheng G, Chen S, Luo W. Synergistically Tuning Water and Hydrogen Binding Abilities Over Co4N by Cr Doping for Exceptional Alkaline Hydrogen Evolution Electrocatalysis. Adv. Energy Mater. 2019;9:1902449. DOI: 10.1002/aenm.201902449

[67] Liu, B.; He, B.; Peng, H.-Q.; Zhao, Y.; Cheng, J.; Xia, J.; Shen, J.; Xiangmin, T.-W. N.; Lee, M. C.-S.; Zhang, W. Nickel Nitride Enriched with Nitrogen Vacancies as a High-Efficiency Electrocatalyst for Hydrogen Evolution. Adv. Sci. 2018;5:1800406. DOI: 10.1002/advs.201800406

[68] Shalom M, Molinari V, Esposito D, Clavel G, Ressnig D, Giordano C, Antonietti M. Sponge-like Nickel and Nickel Nitride Structures for Catalytic Applications. Adv. Mater. 2013;26:1272. DOI: 10.1002/adma.201304288

[69] Zhang B, Xiao C, Xie S, Liang J, Chen X, Tang Y, Iron–Nickel Nitride Nanostructures in Situ Grown on Surface-Redox-Etching Nickel Foam: Efficient and Ultrasustainable Electrocatalysts for Overall Water Splitting. Chem. Mater. 2016;28:6934. DOI: 10.1021/acs.chemmater.6b02610

[70] Jia X, Zhao Y, Chen G, Shang L, Shi R, Kang X, Waterhouse G, Wu L-Z, Tung C-H, Zhang T. Ni3FeN Nanoparticles Derived from Ultrathin NiFe-Layered Double Hydroxide Nanosheets: An Efficient Overall Water Splitting Electrocatalyst. Adv. Energy Mater. 2016;6:1502585. DOI: 10.1002/aenm.201502585

[71] Zhang Y, Ouyang B, Xu J, Chen S, Rawat R-S, Fan H-J. 3D Porous Hierarchical Nickel–Molybdenum Nitrides Synthesized by RF Plasma as Highly Active and Stable Hydrogen-Evolution-Reaction Electrocatalysts. Adv. Energy Mater. 2016;6:1600221. DOI: 10.1002/aenm.201600221

[72] Ma L, Ting L, Molinari V, Giordano C, Yeo B-S. Efficient hydrogen evolution reaction catalyzed by molybdenum carbide and molybdenum nitride nanocatalysts synthesized via the urea glass route. J. Mater. Chem. A. 2015;3:8361. DOI: 10.1039/C5TA00139K

[73] Shalom M, Ressnig D, Yang X, Clavel G, Fellingera T-P, Antoniettia M. Nickel nitride as an efficient electrocatalyst for water splitting. J. Mater. Chem. A. 2015;3:8171. DOI: 10.1039/C5TA00078E

[74] Xing Z, Li Q, Wang D, Yang X, Sun X. Self-supported nickel nitride as an efficient high-performance three-dimensional cathode for the alkaline hydrogen evolution reaction. Electrochem. Acta. 2016;191:841. DOI: 10.1016/j.electacta.2015.12.174

[75] Han L, Feng K, Chen Z. Self-Supported Cobalt Nickel Nitride Nanowires Electrode for Overall Electrochemical Water Splitting. Energy Technol. 2017;5:1908. DOI: 10.1002/ente.201700108

[76] Xue Z, Kang J, Guo D, Zhu C, Li C, Zhang X, Chen Y. Self-supported cobalt nitride porous nanowire arrays as bifunctional electrocatalyst for overall water splitting. Electrochimica Acta. 2018;273:229-238. DOI: 10.1016/j.electacta.2018.04.056

[77] Chen Z, Ha Y, Liu Y, Wang H, Yang H, Xu H, Li Y, Wu R. In Situ

Formation of Cobalt Nitrides/Graphitic Carbon Composites as Efficient Bifunctional Electrocatalysts for Overall Water Splitting. ACS Appl. Mater. Interfaces. 2018;10: 7134. DOI: 10.1021/acsami.7b18858

[78] Zhu C, Wang AL, Xiao W, Chao D, Zhang X, Tiep N, Chen S, Kang J, Wang X, Ding J, Wang J, Zhang H, Fan H-J. In Situ Grown Epitaxial Heterojunction Exhibits High-Performance Electrocatalytic Water Splitting. Adv. Mater. 2018;30:1705516. DOI: 10.1002/adma.201705516

[79] Ledendecker M, Schlott H, Antonietti M, Meyer B, Shalom M. Experimental and Theoretical Assessment of Ni-Based Binary Compounds for the Hydrogen Evolution Reaction. Adv. Mater. 2017; 7:1601735. DOI: 10.1002/aenm.201601735

[80] Peng X, Pi C, Zhang X, Li S, Huo K, Chu PK. Recent progress of transition metal nitrides for efficient electrocatalytic water splitting. Sustainable Energy Fuels, 2019;3: 366-381. DOI: 10.1039/C8SE00525G

[81] Yu L, Song S, McElhenny B, Ding F, Luo D, Yu Y, Chen S, Ren Z. A universal synthesis strategy to make metal nitride electrocatalysts for hydrogen evolution reaction. J. Mater. Chem. A 2019;7: 19728-19732. DOI: 10.1039/C9TA05455C

[82] Chen P, Xu K, Tong Y, Li X, Tao S, Fang Z, Chu W, Wu X, Wu C. Cobalt nitrides as a class of metallic electrocatalysts for the oxygen evolution reaction. Inorg. Chem. Front. 2016;3:236-242. DOI: 10.1039/C5QI00197H

[83] Fan Y, Ida S, Staykov A, Akbay T, Hagiwara H, Matsuda J, Kaneko K, Ishihara T. Ni-Fe Nitride Nanoplates on Nitrogen-Doped Graphene as a Synergistic Catalyst for Reversible Oxygen Evolution Reaction and Rechargeable Zn-Air Battery. Small 2017;23: 1700099. DOI: 10.1002/smll.201700099

[84] H.-P. Guo, B.-Y. Ruan, W.-B. Luo, J. Deng, J.-Z. Wang, H.-K. Liu, S.-X. Dou Ultrathin and Edge-Enriched Holey Nitride Nanosheets as Bifunctional Electrocatalysts for the Oxygen and Hydrogen Evolution Reactions. ACS Catal. 2018;8:9686-9696. DOI: 10.1021/acscatal.8b01821

[85] Zhang Y, Ouyang B, Xu J, Chen S, Rawat RS, Fan HJ. 3D Porous Hierarchical Nickel–Molybdenum Nitrides Synthesized by RF Plasma as Highly Active and Stable Hydrogen-Evolution-Reaction Electrocatalysts Adv. Energy. Mater. 2016;6: 1600221. DOI: 10.1002/aenm.201600221

[86] Jin H, Liu X, Vasileff A, Jiao Y, Zhao Y, Zheng Y, Qiao SZ. Single-Crystal Nitrogen-Rich Two-Dimensional $Mo_5N_6$ Nanosheets for Efficient and Stable Seawater Splitting. ACS Nano 2018;12: 12761-12769. DOI: 10.1021/acsnano.8b07841

Chapter 8

# Characterization, Photoelectric Properties, Electrochemical Performances and Photocatalytic Activity of the $Fe_2O_3/TiO_2$ Heteronanostructure

*Salah Kouass, Hassouna Dhaouadi, Abdelhak Othmani and Fathi Touati*

## Abstract

The $Fe_2O_3/TiO_2$ nanocomposite was synthesized on FTO subtract via hydrothermal method. The crystal structure, morphology, band structure of the heterojunction, behaviors of charge carriers and the redox ability were characterized by XRD, HR-TEM, absorption spectra, PL, cyclic voltammetry and transient photocurrent spectra. The as-prepared $Fe_2O_3/TiO_2$ photocatalysts with distinctive structure and great stability was characterized and investigated for the degradation of methylene blue (MB) dye in aqueous solution. The ability of the photocatalyst for generating reactive oxygen species, including $O_2^-$ and $^{\cdot}OH$ was investigated. It was revealed that the combination of the two oxides ($Fe_2O_3$ and $TiO_2$) nano-heterojunction could enhance the visible response and separate photogenerated charge carriers effectively. Therefore, the remarkable photocatalytic activity of $Fe_2O_3/TiO_2$ nanostructures for MB degradation was ascribed to the enhanced visible light absorption and efficient interfacial transfer of photogenerated electrons from to $Fe_2O_3$ to $TiO_2$ due to the lower energy gap level of $Fe_2O_3/TiO_2$ hybrid heterojunctions as evidenced by the UV–Vis and photoluminescence studies. The decrease of the energy gap level of $Fe_2O_3/TiO_2$ resulted in the inhibition of electron–hole pair recombination for effective spatial charge separation, thus enhancing the photocatalytic reactions. Based on the obtained results, a possible mechanism for the improved photocatalytic performance associated with $Fe_2O_3/TiO_2$ was proposed. The $Fe_2O_3/TiO_2$ nanocomposite has a specific capacity of 82 $F.g^{-1}$ and shows a higher capacitance than $Fe_2O_3$.

**Keywords:** $Fe_2O_3/TiO_2$, methylene blue degradation, heterojunction, holes and superoxide radicals, photocatalyst

## 1. Introduction

The environmental impact caused by the discharge of untreated wastewaters, or even partially treated in sewage stations, is an increasingly worrying problem, considering the damage caused to the environment [1]. In view of this, a great effort has been made to develop new technologies aiming the treatment of persistent substances in the environment such as heterogeneous photocatalysis, electrochemical techniques and photoelectrochemical processes [2–6]. Among these processes, the heterogeneous photocatalysis that belongs to the class of the advanced oxidation processes has proved very effective as it mineralizes the contaminations existing liquid phases. Over the last few decades, research in the photocatalysis area has been focusing on improving electrochemical and photocatalytic materials [7–10]. So, various photocatalysts such as titanium dioxide ($TiO_2$) were used. It is one of the most used photocatalysts given its efficiency in pollutant degradation in waste water, because of its inexpensiveness, hard-soluble and long-term photostability. However, there are two defects limiting the use of $TiO_2$ in the photocatalysis: one, its wide band gap energy that limits its response to visible light and the other is the rapid recombination of photogenerated electron–hole which leads to the decrease of its photocatalytic activity. Therefore, much effort has been devoted to solving these problems. One solution to overcome the defects is to construct heterojunction photocatalysts. In order to construct a heterojunction photocatalyst based on $TiO_2$, the adaptation of energy levels between the two components is determining, that is, the conduction band edge of the narrow band gap semiconductor is higher than that of $TiO_2$.$Fe_2O_3$, as a highly active photocatalyst with a band gap of 2.0 eV [11–13], seems to be a good choice except it has a lower conduction band edge. However, the Fermi level (EF) of $Fe_2O_3$ is lower than that of $TiO_2$. Indeed, Yanqing Cong and al. proved that $Fe_2O_3$ nanoparticles present a stronger photo-response under visible light irradiation in the nanostructured $Fe_2O_3/TiO_2$ nanotube electrodes [14]. $Fe_2O_3/TiO_2$ nanocomposites revealed outstanding photocatalytic activity under visible light and were used as photocatalysts in the degradation of oxytetracycline [15–18]. In this study, we have synthesized by a hydrothermal approach a $TiO_2/F_2O_3$ heterojunction photocatalyst that exhibited excellent performance in many fields.

## 2. Characterization

The XRD analysis of $Fe_2O_3/TiO_2$ indicates the formation of the $TiO_2$ anatase phase in the presence of the $Fe_2O_3$ rhombohedral structure. $Fe_2O_3$ and the $Fe_2O_3/TiO_2$ nanocomposite were recorded using high resolution transmission electron microscope (HR-TEM). As shown in **Figure 1a**, the $Fe_2O_3$ nanorods are well dispersed with an average diameter of 50 nm.

The magnified high resolution TEM image (**Figure 1a**) illustrates that the $Fe_2O_3$ nanorods cover the $TiO_2$ nanoparticles surface. Lattice fringes in the HRTEM image (**Figure 1b**) of the binary hybrid nanocomposite $Fe_2O_3/TiO_2$ could be assigned to a lattice spacing of 2.35 Å nm corresponding to the (103) plane of $TiO_2$, while the lattice spacing of 2.52 Å nm could be indexed to the (110) plane of $Fe_2O_3$.

The band gap ($E_g$) for pure $Fe_2O_3$, $TiO_2$ and the $Fe_2O_3/TiO_2$is determined by extrapolating the absorption edge using the following Equation [19]:

$$(\alpha h\nu)^2 = A(h\nu - E_g) \tag{1}$$

The $E_g$ values are respectively, $E_{g1}$ = 3.1 eV, $E_{g2}$ = 1.93 eV and $E_{g3}$ = 2.6 eV for $TiO_2$, $Fe_2O_3$ and $Fe_2O_3/TiO_2$ junction. It is noted that the presence of $Fe_2O_3$ in the material increases the intensity of the bands and shifting them to higher wavelengths compared to $TiO_2$.

Introducing an appropriate amount of transition metal oxide in the $TiO_2$ matrix is a promising alternative used to modulate the band gap of the as-obtained

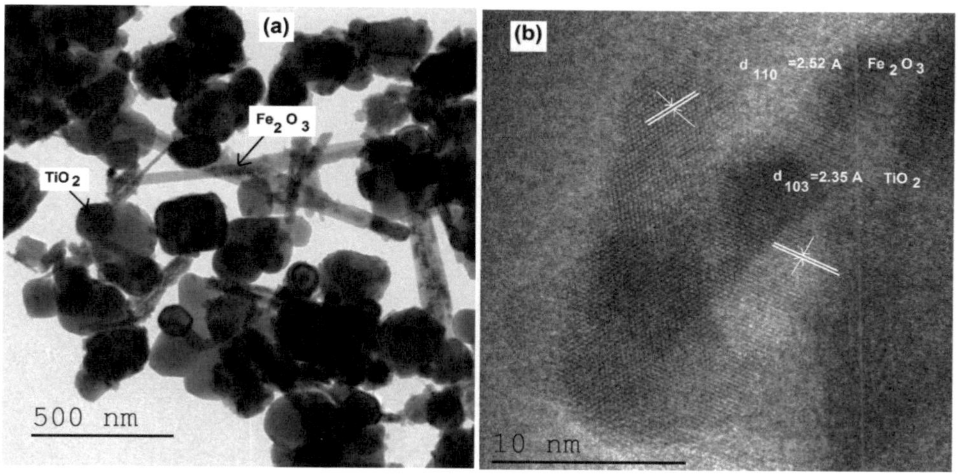


**Figure 1.**
*HRTEM image of: (a) $Fe_2O_3/TiO_2$, (b) HRTEM lattice fringe image of: $Fe_2O_3/TiO_2$.*

nanocomposite. The weakness of band gap due to the electrons excited and injected into the conduction band of $TiO_2$ leads to the improvement of the electron–hole pair separation. The higher degradation activity of the nanocomposite samples ($Fe_2O_3/TiO_2$, CdO/ZnO, $ZnO/TiO_2$, $Bi_2O_3/TiO_2$ and $CdS/TiO_2$) [19–24] is correlated with its lower band gap and strong adsorption in the visible region. In our case, the synthesized $Fe_2O_3/TiO_2$ heterojunction presents a moderate band gap ($E_g$ = 2.6 eV) compared to the other nanocomposite samples. Consequently, the separation efficiency of photogenerated electron–hole pairs in the $Fe_2O_3/TiO_2$ heterojunction could be improved, leading to the improvement of the photocatalytic activity.

## 3. Photocatalytic activity

The study of the photocatalytic activity of $Fe_2O_3/TiO_2$/FTO heterojunction is realized by following the degradation of MB in an aqueous solution under visible light irradiation. The absorption spectrum of MB without a catalyst is characterized by a broad peak centered at 670 nm. The MB absorbance at around 650 nm decreases and there is almost no shift in the peak maximum. This shows that MB was degraded via the destruction of the conjugated structure. This demonstrates that $Fe_2O_3/TiO_2$/FTO exhibits outstanding photocatalytic activity compared to pure $TiO_2$. The obtained results confirm that $TiO_2$ alone is unable to absorb under visible irradiation. This could be attributed to the scaling down of the distance between the valence band (VB) and the conduction band (CB) after the addition of $Fe_2O_3$ which boosts the transfer of electrons between bands after excitation. In comparison, the $Fe_2O_3/TiO_2$/FTO exhibits better photocatalytic degradation than the $TiO_2$ film. Photocatalytic activity is controlled by many factors, such as the phase structure, particle size, light absorption capacity and electron/hole recombination rate [25].

The heterogeneous photocatalysis mechanism has been discussed extensively in the literature [26, 27]. The photo-activity mechanism presented in this study of the $Fe_2O_3/TiO_2$ thin film is described as follows: when the system is irradiated with visible light, the $Fe_2O_3$ electrons at the valence band are excited and hop to the conduction band, leaving a hole ($h^+$). As a result, electron ($e^-$)/hole ($h^+$) pairs are forming. Then, the excited-state electrons produced by $Fe_2O_3$ can be transferred to the conduction band (CB) of the coupled $TiO_2$ due to the existence of electric fields

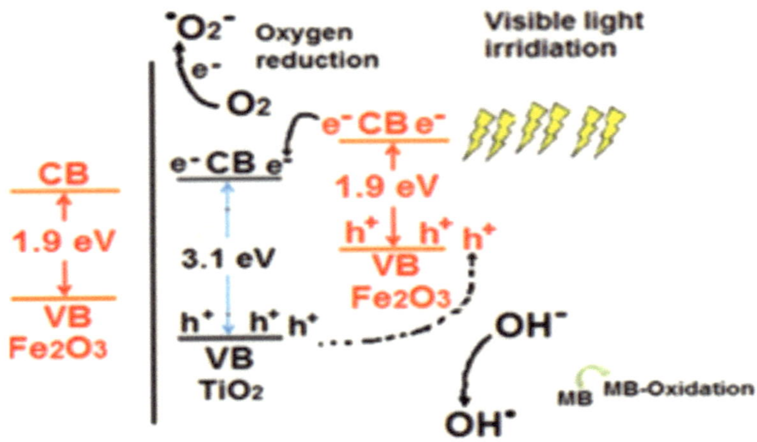


**Figure 2.**
*Schematic illustration of the MB degradation mechanism, under visible light irradiation on the surface of the $Fe_2O_3/TiO_2$/FTO heterojunction.*

between the two materials. At the same time, the photo-generated holes ($h^+$) of $TiO_2$ can quickly transfer to the VB of $Fe_2O_3$. The transferred electrons into the conduction band of $TiO_2$ react with dissolved ($O_2$) to form ($^{\bullet}O_2^-$) and further produce $^{\bullet}OH$. On the other hand, the holes generated on the valence band of $Fe_2O_3$ can easily transfer to that of $TiO_2$ inducing an effective charge separation and transfer. Then the positive charge hole ($h^+$) on $Fe_2O_3$ surfaces reacts with $H_2O$ to generate OH.

**Figure 2** shows the band gap structure and the possible charge carrier transfer between $Fe_2O_3$ and $TiO_2$ under visible light radiation. Before the contact between the two materials, the conduction band (CB) of $TiO_2$ lies above the CB of $\alpha$-$Fe_2O_3$. The energy values of VB ($Fe_2O_3$) and CB ($TiO_2$) were obtained by the two formulas: $E_{VB} = \chi - Ee + 0.5E_g$ and $E_{CB} = E_{VB} - Eg$. [28–30], where $E_g$, $E_e$and $\chi$ represent the band gap energy of the semiconductor, the energy of free electrons (about 4.5 eV) and the electronegativity of the semiconductor, respectively. The $\chi$ values for $TiO_2$ and $Fe_2O_3$ are 5.83 eV and 5.88 eV, respectively [19]. After substituting $\chi$ into the equation, the $E_{VB}$ of $TiO_2$ and $Fe_2O_3$ are found to be 2.93 eV and 2.35 eV, respectively. Therefore, the CB potentials of $TiO_2$ and $Fe_2O_3$ were calculated to be (−0.27 eV) and (0.4 eV), respectively. After coupling $TiO_2$ with $Fe_2O_3$ to form the p-n heterojunction, the Fermi level of $Fe_2O_3$ was dragged upwards, while the Fermi level of $TiO_2$was dragged downwards, until they were at the same level and reached equilibrium [31, 32].

## 4. Photoelectric properties

The PL spectrum is one of efficient approaches to depict the recombination efficiency of photogenerated electron–hole pairs through their different intensities [33]. When the photo-induced electrons and holes are easier to recombine and the lifetime of the photogenerated electrons is shorter, and correspondingly the fluorescence intensity is higher. It is obvious that the PL emission intensity of $Fe_2O_3/TiO_2$ is lower than that of $TiO_2$, implying that the coupling of $TiO_2$ and $Fe_2O_3$ can effectively inhibit the recombination of the photo-generated electron–hole pairs. The decrease recombination rate would be more beneficial for many photocatalyst performance enhancement than $TiO_2$ [34].

**Figure 3** shows the transient photocurrent spectra for $TiO_2$ and $Fe_2O_3/TiO_2$ samples. As shown in **Figure 3**, the maximum photocurrent density of the $Fe_2O_3/TiO_2$/FTO

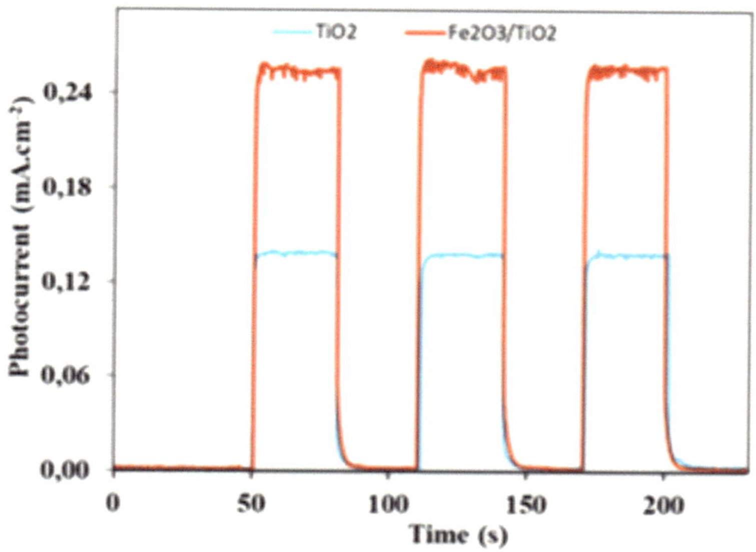


**Figure 3.**
*Transient photocurrent vs. irradiation time for pure $TiO_2$ and $Fe_2O_3/TiO_2$ heterojunction samples in 0.1 M NaOH solution under visible light irradiation.*

heterojunction electrode reached 0.25 $mA.cm^{-2}$ which is almost the twice of the pure $TiO_2$ electrode. This indicates more charge carriers are generated and transferred from conduction band of the photocatalysts to the electrodes. It could be seen that the photocurrent density produced instantly and increased sharply when exposed to visible light, but promptly reduced to zero as soon as the light source is turned off. Those results confirm that $Fe_2O_3/TiO_2$ heterojunction is more effective in generating and separating the photogenerated charge carriers than $TiO_2$, and much faster interfacial charge transfer, benefitting photocatalytic activity.

The improved photocurrent shows the effective interfacial charge transfer between $Fe_2O_3$ and $TiO_2$. Thus, the stable and enhanced photocurrent of the obtained heterojunction is more favorable for photocatalytic dye degradation; furthermore, it is more efficient for water splitting [18].

From the resulting Mott-Schottky plot ($C^{-2}$ versus the applied potential E) [35], the flat band potential $E_{FB}$ could be obtained as the intercept with the $x$-axis and from the slope of the linear part. The value of the flat band potential determined from capacitance measurements is reported to be −0.45 V and − 0.55 V vs. Ag/AgCl for $TiO_2$ and $Fe_2O_3/TiO_2$ Respectively. It should be noted that the $E_{FB}$ of the $Fe_2O_3/TiO_2$ photoanode exhibits a positive shift in comparison to that of the pure $TiO_2$ electrode. Combined with the $E_g$ calculated from the DRS spectra, the optical band gaps of $TiO_2$ and $Fe_2O_3/TiO_2$ are 3.1 and 2.6 eV, respectively. According to the formula $E_g = E_{VB} - E_{CB}$, the valence band positions ($E_{VB}$) of $TiO_2$ and $Fe_2O_3/TiO_2$ are 2.65 and 2.05 V vs. Ag/AgCl, respectively.

## 5. Electrochemical performances

Transition metal oxides such as oxides of Fe, Cu, Ni, Mn, Cu, and TiO2 for electrode materials offer rich redox reactions such as in electrochemical cells providing high specific capacitance values for supercapacitors [36, 37]. Among these metal oxides, $Fe_2O_3$, $TiO_2$, metal doped $TiO_2$ and composite $Fe_2O_3$/ $TiO_2$ are very promising electrodes materials, due to their acceptable charge/discharge capacities [38–40]. Cycling stability and specific capacitance are critical factors in evaluating the electrochemical properties, influenced by synthesis method, morphology and grains size.

The specific capacitance of the electrode calculated from the CV curves, according to the following Equation [40–42]:

$$C = \frac{\int I(V)d(V)}{m\nu\Delta V} \tag{2}$$

α -$Fe_2O_3$ with rod-like structure is synthesized to evaluate as electrode material comparing with Cu foil and Ni foam, the as-prepared electrodes with Ni-foam exhibited higher capacity of 415 mAh $g^{-1}$ and more stable cycle performance [43]. Fe-based materials,$Fe_2O_3$, $Fe_3O_4$, and FeOOH, were synthesized via the microwave–hydrothermal process by Young Dong Noh and al, the results showed that FeOOH had better anode capacity as lithium-ion batteries than those of $Fe_2O_3$ and $Fe_3O_4$ [44]. Yudai Huang and al prepared -$Fe_2O_3$/MWCNTs composites by a simple hydrothermal process and show that initial discharge capacity of $Fe_2O_3$ is 992.3 mAh $g^{-1}$ and the discharge capacity is 146.6 mAh $g^{-1}$ after 100 cycles [45].

## 5.1 $TiO_2$

Anantha Kumar and al demonstrated that synthesis of a grapheme-$TiO_2$ using a microwave technique exhibited a high specific capacitance of 165 F $g^{-1}$at a scan rate of 5 mV $s^{-1}$ [46]. G. Wang and al fabricate $TiO_2$-B nanotubes via a mixed solvothermal technical and subsequent heat treatment and found that specific capacitance is equal to17.7 F/g [47]. The same, capacitances of the CNTs, CNTs/TiO2 composite and UVlight irradiated CNTs/TiO2 composite materials were 4.1F/g; 6.4F/g and 9.8F/g, respectively [48].

## 5.2 TiO2-Fe2O3 composite

The performances of $TiO_2$-$Fe_2O_3$ composite prepared using abundant ilmenite via a heat treatment are improved compared with that of P25, with the increased iron oxide content, the capacity gets higher [49]. Again, theα -$Fe_2O_3$/$TiO_2$/C composite fibers prepared by Luis Zuniga and al, via centrifugal spinning and subsequent thermal processing, showed a superior specific capacity of 340 mAh $g^{-1}$after 100 cycles, compared to 61 mAh $g^{-1}$and 121 mAh $g^{-1}$for $TiO_2$/C and α -$Fe_2O_3$/C materials, respectively [50]. So, $TiO_2$/$FeTiO_3$@C porous materials, synthesized by carbonizing the mixture of pyrrole with lab-made $TiO_2$/ $Fe_2O_3$, have a superior capacity of 441.5 mAh $g^{-1}$after 300 cycles, Comparing with $TiO_2$, $TiO_2$@C, and $TiO_2$/ $Fe_2O_3$ [51]. Also, $TiO_2$, $Fe_2O_3$ NPs and $TiO_2$- $Fe_2O_3$ are synthesized via green combustion method with *Aloe Vera* gel as a fuel. Compared to pure $TiO_2$ and $Fe_2O_3$ materials, the composite showed stable electrochemical performance after 1000 cycles, which can be beneficial for rechargeable supercapacitor [52].

Based on the synergy between the two metallic oxides, $TiO_2$ and $Fe_2O_3$, we have produced $Fe_2O_3$/$TiO_2$ nanocomposite and heterojunction film via the hydrothermal process. The electrochemical performance of the $Fe_2O_3$ and $Fe_2O_3$/$TiO_2$ nanostructures as potential electrode material for supercapacitors, cyclic voltammetry (CV) tests were performed in a three-electrode cell with $Na_2SO_4$ aqueous electrolyte. The current response as a function of the potential applied to the working electrode, recorded at 100 mV/s in a potential range between −0.8 V and − 0.4 V with 1000 cycles of the $Fe_2O_3$/$TiO_2$ nanocomposite in the

$Na^+$-system. Therefore, the $Fe_2O_3/TiO_2$ nanocomposite shows electrochemical and structural stability during redox cycling.

The excellent pseudocapacitive performance of the $Fe_2O_3/TiO_2$ nanocomposite electrode is probably attributed to the positive synergistic effects between the $Fe_2O_3$ and $TiO_2$.

First, This combination can not only inhibit the agglomerating of $TiO_2$ nanoparticles but also reduce the aggregation of the nanoparticles made the nearly every Fe2O3 nanoparticle access to the electronic and ionic transport pathways resulting in high double–layer capacitance, and importantly, enhancing the utilization of active materials [28]. Second, the large distance between neighboring graphene nanosheets provide enough void spaces to buffer volume change during the redox reaction, and endow good electrical contact with the nanoparticles upon cycling [32, 37]. Third, the unique structure can facilitate the diffusion and migration of the electrolyte ions that can increase the specific capacitance value and improve the high rate charge–discharge performance [38, 39]. Finally, graphene also provides a highly conductive network for electron transport during the charge and discharge processes, thus reducing the polarization of the electrodes [44–46].

## 6. Conclusion

$TiO_2$, $Fe_2O_3$ hematite nanoparticles and $Fe_2O_3/TiO_2$ nanocomposites were synthesized via a simple hydrothermal process. The photocatalytic activity of the $Fe_2O_3/TiO_2$ nanocomposite was evaluated using the degradation of methylene blue (MB) under sunlight irradiation for pollution prevention. The results proved that the $Fe_2O_3/TiO_2$heterojunction has a higher removal efficiency of MB and stronger photo-response under visible light irradiation. Compared to both pure $TiO_2$ and $Fe_2O_3$, the $Fe_2O_3/TiO_2$photocatalyst have enhanced photocatalytic activity. This improved activity of the heterojunction between the $TiO_2$ and $Fe_2O_3$ nanoparticles results from the improved charge transfer and suppressed electron–hole recombination. We have also compared the photoelectric properties of $Fe_2O_3/TiO_2$heterogeneous photocatalysts with that of pure $TiO_2$. The obtained result demonstrated that the formation of heterojunction between $Fe_2O_3$ and $TiO_2$ was pivotal for improving the separation and thus restraining the recombination of photogenerated electrons and holes, which accounts for the enhancement of photocatalytic activity. The study of the role of the active species on $Fe_2O_3/TiO_2$ confirmed that the crucial active species were both holes and superoxide radicals. The $Fe_2O_3/TiO_2$sample also showed good stability and reusability, suggesting its potential for water purification applications. Likewise, the visible photogenerated electrons in the obtained heterojunction would provide a feasible route to improve solar water splitting, which will be investigated in further studies. The electrochemical properties of the as-synthesized nanocomposite materials ($\alpha$-$Fe_2O_3/TiO_2$) were evaluated by cyclic voltammetry for 1000 cycles. The $\alpha$-$Fe_2O_3/TiO_2$ nanocomposite materials exhibited an enhanced specific discharge capacity compared to $Fe_2O_3$ nanomaterials. The as-fabricated hybrid electrodes show an impressive performance as a high-capacity anode for $Na^+$-ion batteries.

## Author details

Salah Kouass[1]*, Hassouna Dhaouadi[2], Abdelhak Othmani[3] and Fathi Touati[2]

1 Laboratoire des Matériaux Utiles, INRAP, Technopole Sidi-Thabet, Ariana Tunis, Tunisie

2 Laboratoire Matériaux, Traitement et Analyses, INRAP, Technopole Sidi-Thabet, Ariana Tunis, Tunisie

3 Laboratoire de Physique des Matériaux: Structure et Propriétés, Faculté des Sciences de Bizerte, Université de Carthage, Bizerte, Tunisia

*Address all correspondence to: koissa2000@yahoo.fr

## References

[1] Shahmoradi, B.; Ibrahim, I. A.; Sakamoto, N.; Ananda, S.; Row, T. N. G.; Soga, K.; Byrappa, K.; Parsons, S.; Shimizu, Y.;Dye Degradation Enhanced by Coupling Electrochemical Process and Heterogeneous Photocatalysis, Environ. Technol. 31(2010) 1213.

[2] Benehkohal, N. P.; Demopoulos, G. P.;"Electrophoretically self-assembled mixed oxide-$TiO_2$ nano-composite film structures for photoelectrochemical energy conversion : Probing of charge recombination and electron transport resistances, J. Power Sources, 240 (2013) 667.

[3] Calva-Yanez, J. C.; Rincon, M. E.; de la Fuente, M. S.; Alvarado- Tenorio, G.Structural and photoelectrochemical characterization of MWCNT-TiO2 matrices sensitized with $Bi_2S_3$;J. Solid State Electrochem. 17 (2013) 2633.

[4] Khataee, A. R.; Zarei, M.; Ordikhani-Seyedlar, R.; Heterogeneous photocatalysis of a dye solution using supported $TiO_2$nanoparticles combined with homogeneous photoelectrochemical process:Molecular degradation products, *J. Mol. Catal. A: Chem. 338*(2011)84.

[5] Ochiai, T.; Fujishima, A.;Photo-electrochemical properties of $TiO_2$ photocatalyst and its applications for environmental purificationJ. Photo-chem. Photobiol., C, 13 (2012) 247.

[6] Zhang, Z.; Wang P.,Optimization of photoelectrochemical water splitting performance on hierarchical $TiO_2$ nanotube arrays; Energy Environ. Sci., 5 (2012)6506.

[7] Jinlin Long, Hongjin Chang, QuanGu, Jie Xu, Lizhou Fan, Shuchao Wang, Yangen Zhou, Wei Wei, Ling Huang, Xuxu Wang, Ping Liu, and Wei Huang, Gold-plasmon enhanced solar-to-hydrogen conversion on the {001} facets of anatase $TiO_2$nanosheets, Energ. & Environ. Sci., 7(2014) 973-977.

[8] LingshuMeng, Zhenye Chen, Zhiyun Ma, Sha He, YidongHou, Hao-Hong Li, Rusheng Yuan, Xi-He Huang, Xuxu Wang, Xinchen Wang, Jinlin Long, Gold Plasmon-Induced Photocatalytic Dehydrogenative Coupling of Methane to Ethane on Polar Oxide Surfaces, Energ. & Environ. Sci. 11 (2018) 294-298.

[9] Jie Xu, Liufeng Luo, Guangrui Xiao, Zizhong Zhang, Huaxiang Lin, Xuxu Wang, and Jinlin Long, Layered $C_3N_3S_3$ Polymer/Graphene Hybrids as Metal-Free Catalysts for Selective Photocatalytic Oxidation of Benzylic Alcohols under Visible Light, ACS Catal. 4 (2014) 3302-3306.

[10] C. C. Wang, Y. Zhan and Z. Y. Wang,$TiO_2$, $MoS_2$, and $TiO_2/MoS_2$ Heterostructures for Use in Organic Dyes Degradation, ChemistrySelect, 3 (2018) 1713-1718.

[11] Hongwen Zhang, Lei Ma, Jintao Ming, Bingqian Liu, YiboZhao, YidongHou, Zhengxin Ding, Chao Xu, Zizhong Zhang, Jinlin Long, Amorphous $Ta_2O_xN_y$-Enwrapped $TiO_2$ Rutile Nanorods for Enhanced Solar Photoelectrochemical Water Splitting, Applied Catalysis B: Environmental 243 (2019) 481-489.

[12] R.L. Spray, K.J. McDonald, K.S. Choi, Enhancing photoresponse of nanoparticulate alpha-$Fe_2O_3$ electrodes by surface composition tuning, J. Phys. Chem C. 115 (8) (2011) 3497-3506.

[13] A. Umar, M.Abaker, M. Faisal, S.W. Hwang, S.Baskoutas, S.A. Al-Sayari, High-yield synthesis of well-crystalline alpha-$Fe_2O_3$ nanoparticles: structural, optical and photocatalytic properties,J. Nanosci. Nanotechnol. 11 (4) (2011) 3474-3480.

[14] Y. Cong, Z. Li, Z. Yi, W. Qi, X. Qian, Synthesis of $\alpha$-$Fe_2O_3/TiO_2$ nanotube arrays for photoelectro-

Fenton degradation of phenol,Chem. Engin. J. 191(2012) 356-363.

[15] L.Rong, J.Yuefa, W. Jun, Z.Qiang, Photocatalytic Degradation and Pathway of Oxytetracycline in Aqueous Solution by $Fe_2O_3/TiO_2$, Nanopowders,R SCAdvances.51(2015) 1-9.

[16] M. Ni, M. K. H. Leung, D.Y.C. Leung, K. Sumathy, A review and recent developments in photocatalytic water-splitting using $TiO_2$ for hydrogen production, Renew. Sustain. Energy Rev. 11 (2007) 401-425

[17] J. Tang, J.R. Durrant, D.R. Klug, Mechanism of photocatalytic water splitting in $TiO_2$ reaction of water with photoholes, importance of charge carrier dynamics, and evidence for four-hole chemistry, J. Am. Chem. Soc. 42 (2008) 13885-13891.

[18] P. Luan, M. Xie, X. Fu, Y. Qua, X. Sun, L. Jing, Improved photoactivities of $TiO_2/Fe_2O_3$ nanocomposites for visible-light water splitting after phosphate bridging and mechanism, Phys. Chem. Chem. Phys. 17 (2015) 5043-5050.

[19] C. Wei, Z.Maojin, R.Xiaosai, M. Andrew, Synthesis and Photocatalytic Activity of Monolithic $Fe_2O_3/TiO_2$, S. Afr. J. Chem. 70(2017) 127-131.

[20] D. Lee, Y. Rho, F. Allen, *A. minor*, S.H. Ko and C. Grigoropoulos, Synthesis of Hierarchical $TiO_2$ Nanowires with Densely-Packed and Omnidirectional Branches, Nanoscale. (2013)

[21] A Othmani, S. Kouass, Th. Khalfi, S. Bourchada, F. Touati and H. Dhaouadi, Studies of the photocatalytic and electrochemical performance of the $Fe_2O_3/TiO_2$ heteronanostructure, Journal of the Iranian Chemical Society, https://doi.org/10.1007/s13738-020-01993-0.

[22] J. Tauc, R.Grigorovici, A.Vancu, Optical Properties and Electronic Structure of Amorphous Germanium, Phys. Stat. Sol. 15(1966) 627-637.

[23] S. Yang, X.Quan, X. Li, Y. Liu, S.Chen, G. Chen, Preparation, characterization and photo-electrocatalytic properties of nanocrystalline $Fe_2O_3/TiO_2$, ZnO/ $TiO_2$and $Fe_2O_3$/ZnO/$TiO_2$ composite film electrodes towards pentachloro-phenol degradation, Phys ChemChem Phys. 6 (2004) 659- 664.

[24] P.Linlin, X.Tengfeng, L.Yongchun, F.Haimei,W.Dejun, Synthesis, photoelectric properties and photocatalytic activity of the $Fe_2O_3/TiO_2$ heterogeneous photocatalysts, Phys. Chem. Chem. Phys.12(2010) 8033-8041.

[25] L.Yin, Y. Xia, Core-Shell Structured $\alpha$-$Fe_2O_3$@$TiO_2$ Nanocomposites with Improved Photocatalytic Activity in Visible Light Region, Phys. Chem. Chem. Phys. 42(2013) 1-7

[26] J. Aslam, M.R. Mohammed, F. Mohammed, B.K. Sher, Studies on Photocatalytic Degradation of Acridine Orange and Chloroform Sensing Using as-Grown Antimony Oxide, Micro-structures. Mater. Sc. Appl. 2(2011) 676-683

[27] L.P. Zhu, N.C. Bing, D.D. Yang, Y. Yang, G.H. Liao, L.J Wang, Synthesis and photocatalytic properties of core–shell structured $\alpha$-$Fe_2O_3$@$SnO_2$shuttle-like nanocomposites, Cryst. Eng. Comm. 13 (2011) 4486-4490.

[28] M.R. Mohammed, M.A. Abdullah, E.Y. Tamer, M.M.Hadi, Photocatalytic degradation of remazol brilliant orange 3R using wet-chemically prepared CdO-ZnO nanofibers for environmental remediation, Mater. Express. 6(2016) 137-148.

[29] W.Liugang, Z.Junying, L.Chunzhi, Z. Hailing, W.Wenwen,W.Tianmin, Synthesis, Characterization and Photocatalytic Activity of $TiO_2$ Film/

$Bi_2O_3$Microgrid Heterojunction, J Mater Sci Technol. 27(2011) 59-63.

[30] Z.Panpan, L.Zhanggao, X. Yu, F. Jing, X.Jiangwei, Studies on facile synthesis and properties of mesoporous CdS/$TiO_2$ composite for photocatalysis applications, J. Alloys. Compounds. 692(2017) 170-177.

[31] L. Peng, T. Xie, Y. Lu, H. Fan, D. Wang, Synthesis, Photoelectric Properties and Photocatalytic Activity of the $Fe_2O_3$/$TiO_2$ Heterogeneous Photocatalysts, Phys. Chem. Chem. Phys. 12(2010) 8033-8041.

[32] M.Niu, F. Huang, L. Cui, P. Huang, Y. Yu, Y.Wang,Hydrothermal Synthesis, Structural Characteristics, and Enhanced Photocatalysis of $SnO_2$/α-$Fe_2O_3$ Semiconductor Nanohetero-structures,ACS.Nano. 4(2010) 681-688.

[33] Cong Y., Zhang J., Chen F., Anpo, M. J. Phys. Chem. C, 111 (2007) 6976-6982.

[34] S.D. Deldkar, H.M. Yadav, S.N. Achary, S.S. Meena and S.H. Pawar, Structural refinement and photocatalytic activity of Fe-doped anatase $TiO_2$ nanoparticles, Appl. Surf. Sci. 263 (2012) 536-545.

[35] Z. Chen, H. N. Dinh and E. Miller, Photoelectrochemical water splitting, Springer Briefs in Energy, New York, (2013) 49-61.

[36] X. Zhang, P. Suresh Kumar, V. Aravindan, H.H. Liu, J. Sundaramurthy, S.G. Mhaisalkar, H.M. Duong, S. Ramakrishna, S. Madhavi , Electrospun $TiO_2$—Graphene Composite Nanofibers as a Highly Durable Insertion Anode for Lithium Ion Batteries, J. Phys. Chem. C 116 (2012) 14780-14788.

[37] L. Zuniga, V. Agubra, D. Flores,; H. Campos J. Villareal, M. Alcoutlabi, Multichannel hollow structure for improved electrochemical performance of $TiO_2$/Carbon composite nanofibers as anodes for lithium ion batteries, J. Alloys Compd. 686(2016) 733-743.

[38] L.W. Ji, Z. Lin, M. Alcoutlabi, X.W. Zhang, Recent developments in nanostructured anode materials for rechargeable lithium-ion batteries, Energy Environ. Sci. 4(2011) 2682-2699.

[39] L. Gao, H. Hu, G.J. Li, Q.C. Zhu, Y. Yu, Hierarchical 3D $TiO_2$@$Fe_2O_3$ nanoframework arrays as high-performance anode materials, Nanoscale 6(2014) 6463-6467.

[40] S. Li, M.Y. Wang, Y. Luo, J.G. Huang, Bio-Inspired Hierarchical Nanofibrous $Fe_3O_4$-$TiO_2$-Carbon Composite as a High-Performance Anode Material for Lithium-Ion Batteries. ACS Appl. Mater, Interfaces, 8(2016) 17343-17351.

[41] S.S. Shinde, G.S. Gunda, D.P. Dubal, S.B. Jambure, C.D. Lokhande, Morpho-logical modulation of polypyrrole thin films through oxidizing agents and their concurrent effect on supercapacitor performance,Electrochim.Acta.119 (2014) 1-10.

[42] D. Qu, H. Shi, Studies of activated carbons used in double-layer capacitors, J. Power. Sources. 74 (1998) 99-107.

[43] L. Huang, Z. Min, Q. Zhang, Research and analysis on electro-chemical performances of α -$Fe_2O_3$ electrode in Li-ion battery with different current collectors, Materials Research Bulletin 66 (2015) 39-44.

[44] K. Chen, Y. D. Noh, W. Huang, J. Ma, SridharKomarneni, Dongfeng Xue, Microwave-hydrothermal synthesis of Fe-based materials for lithium-ion batteries and supercapacitors. Ceramics International 40 (2014) 2877-2884.

[45] Y. Huang, Z. Dong, D. Jia, Z. Guo d, Won Il Cho, Electrochemical properties of α$Fe_2O_3$/MWCNTs as anode materials for lithium-ion batteries, Solid State Ionics 201 (2011) 54-59.

[46] A. Ramadoss, Sang Jae Kim, Improved activity of a grapheme-$TiO_2$ hybrid electrode in an electrochemical supercapacitor, CA R B O N 6 3 (2013) 434-445.

[47] G. Wang, Z.Y. Liu, J.N. Wu, Q. Lu, Preparation and electrochemical capacitance behavior of $TiO_2$-B nanotubes for hybrid supercapacitor, Materials Letters 71 (2012) 120 122.

[48] Bin Zhang, Rui Shi, Yupeng Zhang, Chunxu Pan, CNTs/$TiO_2$ composites and its electrochemical properties after UVlight irradiation, Progress in Natural Science: Materials International 23(2) (2013)164-169.

[49] Li. Li, J. Zhangb and Q. Zhua, Novel fractional crystallization route to porous $TiO_2$-$Fe_2O_3$ composites: large scale preparation and high performances as photocatalyst and Li-ion battery anode, Dalton Trans. 2015, DOI: 10.1039/ C5DT04091D.

[50] Luis Zuniga, Gabriel Gonzalez, Roberto Orrostieta Chavez, Jason C. Myers,Timothy P. Lodge and Mataz Alcoutlabi, Centrifugally Spunα -$Fe_2O_3$/ $TiO_2$/Carbon Composite Fibers as Anode Materials for Lithium-Ion Batteries, Appl. Sci. 9 (2019) 4032.

[51] S. Guoa, J. Liu, S. Qiu, Y. Wang, X. Yan, N. Wu, S. Wang, Z. Guo, Enhancing Electrochemical Performances of $TiO_2$ Porous Microspheres through Hybridizing with $FeTiO_3$ and Nano-carbon, Electrochimica Acta 190 (2016) 556-565.

[52] M. R. Anil Kumar, B. Abebe, H. P. Nagaswarupa, H. C. Ananda Murthy, C. R. Ravikumar & Fedlu Kedir Sabir, Enhanced photocatalytic and electro-chemical performance of $TiO_2$-$Fe_2O_3$ nanocomposite: Its applications in dye decolorization and as supercapacitors, Scientific Reports 10 (2020) 1249.